On a field in central Utah in 1947 a farmer and his son
are harrowing and planting. Mechanized power was well
underway toward replacing muscle and horsepower.

Hogs are raised in all fifty states, but mostly in the Midwest near the corn and soybeans that make up the bulk of their food. Like this Wisconsin hog raiser in 1951, the farmer today may still depend upon his own experience and that of his neighbors, but he also makes decisions on the basis of information he gets from the Department of Agriculture, land grant colleges and universities, farm magazines and major suppliers of seeds, chemicals and machinery.

James Jepson, 1947, then ninety-three and legally blind, was the first president of the Hurricane Canal Company, Hurricane, Utah. This farmer's imagination led to the project that sluiced Virgin River water eight miles from Zion National Park to the arid terrain of the Hurricane Bench in southern Utah. The rough-hewn water system took 10 years to build, using trestles over gorges and tunneling through the mountains, transforming 2,000 acres of desert into fertile farmland.

CONTENTS

Pig at feed trough, Indiana, 1955. As a result of improvement in productivity, consumer food costs have risen less than any other necessity item in the cost-of-living index. Farmers, however, receive less for their contribution to the "farm food market basket" than they did at the end of World War II. They get about 76 cents for the pork used in chops costing $2.00 at the butcher counter and 9 cents for the corn in a box of breakfast flakes priced at a $1.50. In the face of this declining food dollar share, only increased productivity has enabled them to stay in business.

PREFACE

Between the Divining Rod and the Computer

*T*he changes in agriculture, even within our memory, urge us to record and preserve that which was. New machines, new seeds, new chemicals, new livestock breeding, and new dimensions in farm management had been researched and partially developed as World War II came to a close; but it was the following three decades, perhaps more than any other comparable period, that saw the great implementation of these "revolutions." Since the mid-1970s there has been some increase in yields and some new developments, particularly in the area of biotechnology, but nothing so far to compare in quantity and substance to the period from the mid-1940s to the mid-1970s.

Arable acreage, of course, remained the

same; but the number of working farm units was declining. In my lifetime the number of farms has decreased from around 6 1/2 million in 1917 to less than 2 million as we enter the 1990s, while the number of people each farmer feeds has increased from 5 to 6 to close to 100.

It was a time of transition between old and new as the work ethic and social values of the family farm were mixing with on-rushing technology and agribusiness consolidation. Coinciding with this was the enormous increase in food production needed to feed the armies of WWII and the hungry populations of a war-devastated world.

All this stimulated bountiful harvests from American farmers, and sparked a new vision of even greater production. New technology made it possible for farmers to increase yields in quantum leaps by the 1950s. Farmers have always willingly embraced new ways to make farming more profitable and satisfying. What was new was the scale of increased production, and the rapidity of change during those post-war years.

Despite these great increases in productivity, farmers are still prey to unique uncertanties: diseases and insects can decimate their livestock or crops; drought, flooding, wind or hail can ravage their land, and a superb crop can paradoxically cause prices to tumble below costs of production. In spite of the best efforts of scientists, engineers and economists, the farmer is more vulnerable than most businessmen to situations, both domestic and foreign, beyond his control. Much of the public is unaware of these problems.

By 1990, some fifteen-years after the last of these photographs was taken, there are still persistent economic aberrations. Many regions of the world that were formerly markets for the U.S. farmer are now either largely self-sufficient, or, in some cases, have actually become exporters. Today the American farmer has practically no control over the prices he gets for his product, yet he pays the posted price for his supplies and machinery. In many cases, he receives less for his product today than he did at the end of World War II. On a contrary treadmill he battles for efficiency by continually enhancing his yields, which in turn adds to surpluses and further depresses market prices.

He wants the government "off his back," yet admits that the United States Department of Agriculture is the only entity with the size, ability and geographic reach to coordinate information on all the variables of agriculture, and formulate programs to cushion the pricing swings of the market place. Consumer food costs have risen less than any other necessity in the cost-of-living index.

During those three decades, I was roaming the U.S. on photographic assignments for farming and general interest publications, as well as government agencies and corporations. Even though I handled many kinds of journalistic subjects, I developed a strong affection for the land and the people who farmed it, and became something of a specialist in interpreting farm subjects. Yet I did not realize at the time that this period was to be in any way historically unique, and my working perception was not at all oriented to recording any particular strand of agricultural history.

This collection is a harvest from many hundreds of images from that period—my photographic impression of what the farmer does, sees, and feels. There seems to be reflected here an image of those changing times on the American farm; changes that are interwoven with the two constants of agriculture—the farmer's love for the land and working with living things.

In a world replete with multiple means of destruction, the farmer's bright business is creating and sustaining life.

Joe Munroe, 1990

Orval Proud, Ray Smith and Dr. Ralph Wadley were three of the farmers on the "Mile of Farms" in Michigan in 1947 when Joe Munroe produced a story on the families who farmed there. See page 42 for an update on the "Mile of Farms."

INTRODUCTION

Those Good Old Days

In 1946, the year Joe Munroe began his career in photo journalism, my father made a major career decision of his own. He traded Hank, John, Queen and Pat—his Belgian workhorses—even up at O'Kones Implement Company, Wellsburg, Iowa, for a brand-new Farmall H.

"For $150 more I could have bought an M [a more powerful tractor]," John H. Kruse, who lived 50 years on the same farm in Grundy County, recalled 44 years later. "The H was small and slow, but at the time I just couldn't spend another $150. That was a lot of money in those days."

While Dad eagerly anticipated the labor-saving benefits of mechanical farming, my mother was less ready to welcome the shiny red H as a replacement for the dapples and sorrels. "I cried when the horses left the farm for the last time," remembers Johanna Kruse. "But I guess trading for the tractor was the best thing to do."

I agree. Of course that's easy for me to say, for I wasn't born until the year after the H arrived in the yard. All I knew about horses was horsepower—the kind found in engines. Tractors became increasingly more sophisticated as well as muscular as the 50s, 60s and 70s yielded a bumper crop of new technology and ideas.

It was a great time to be a boy growing up on the farm. I'm not one to talk about "those good old days," but indeed, for me those *were* good days.

Joe Munroe and his camera were there on the farms of America to capture the tremendous changes of the three decades following World War II. His photo images convey powerful feelings and emotion. The faces of the people, livestock and the land reflect the story of the times as much as the new machines and farming practices he photographed. Those of you who were living on farms in that era will see yourselves in the people and situations depicted in the book's 200 photographs.

As much as the tools of farming change, the love of farmers for the land and care for livestock remain unchanged. One observer has said that farming has always been about rocks, weeds and flies. Rocks in the broadest sense are the collection of agrono-

mic factors which determine yield. Weeds will always be weeds. Flies represent the irritations and diseases of livestock. That

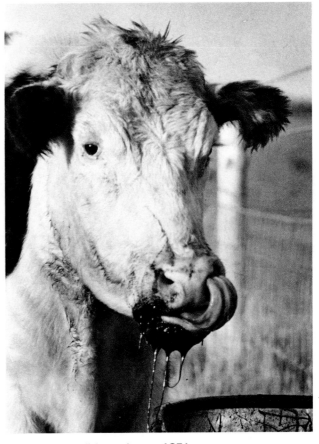

An Iowa steer licks molasses, 1956.

pretty well sums up the production challenges of farming ever since Adam and Eve were banished from the Garden of Eden to tend the thorny fields.

Whether you relive "those good old days" in this book or are introduced to them for the first time, I know your understanding and appreciation of farmers will be enriched. ■

Loren Kruse, Editor
***Successful Farming* magazine**

Penned-up sheep give a leap when first released. Judd McKnight sheep ranch near Roswell, New Mexico, 1957.

Farm road next to ladino clover pasture, central Ohio, 1947.

South-central Wisconsin farmstead in 1947 with fields plowed on the contour to lessen soil erosion.

*I*n 1946 there were over 5 million farms; in 1975 with roughly the same total acreage there were 2.8 million, but the rate of decline was slowing. There was a loss of 22,000 farm units in 1975; in 1950 the loss figure was 220,000. The average size of a U.S. farm in 1975 was 385 acres.

California central valley farmstead near Fresno, 1966.
Western farms are generally larger, and on flat land. With
much less annual rainfall they are also more dependent on
irrigation systems using publicly funded water-delivery
sources from nearby mountains.

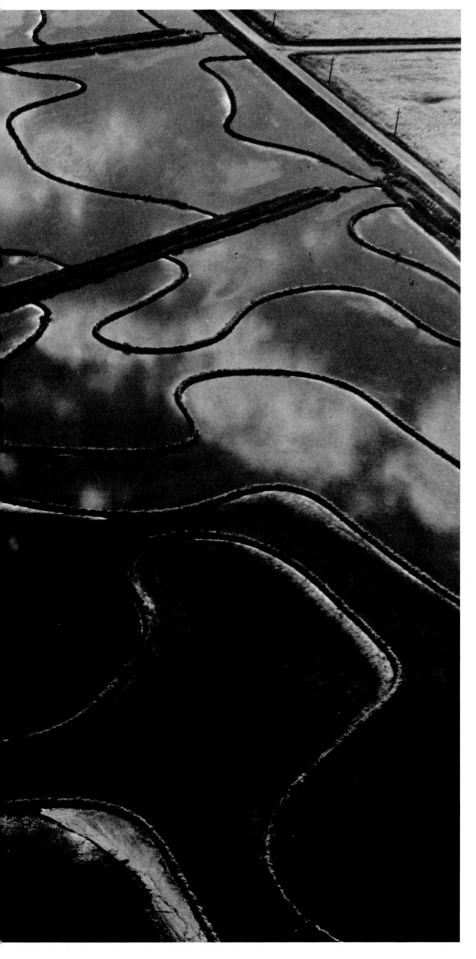

Rice fields, Colusa, California 1961.

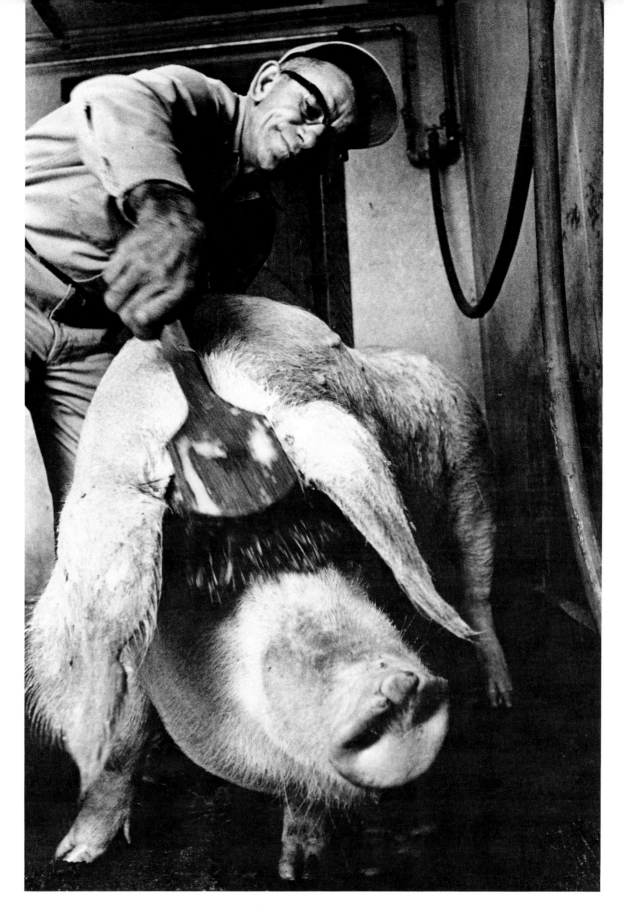

Washing a pig, Missouri, 1962. Hogs are sometimes
known as "mortgage lifters" because they probably have
paid for more farms than any other single farm product.

◀ Suckling piglets, Missouri, 1960.

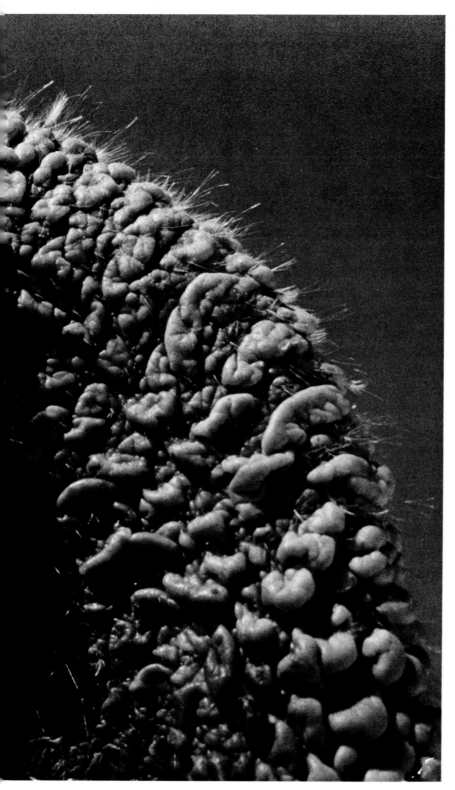

Turkey, Missouri, 1962. Mass production and new technology came to turkey raising in the post-war years. Large commercial turkey farms began breeding for various sizes of birds, and meat distribution on the bird. Artificial insemination crews could inseminate about 700 hens an hour at a cost of about $.06 per bird. Even today, no one knows why a turkey has a wattle.

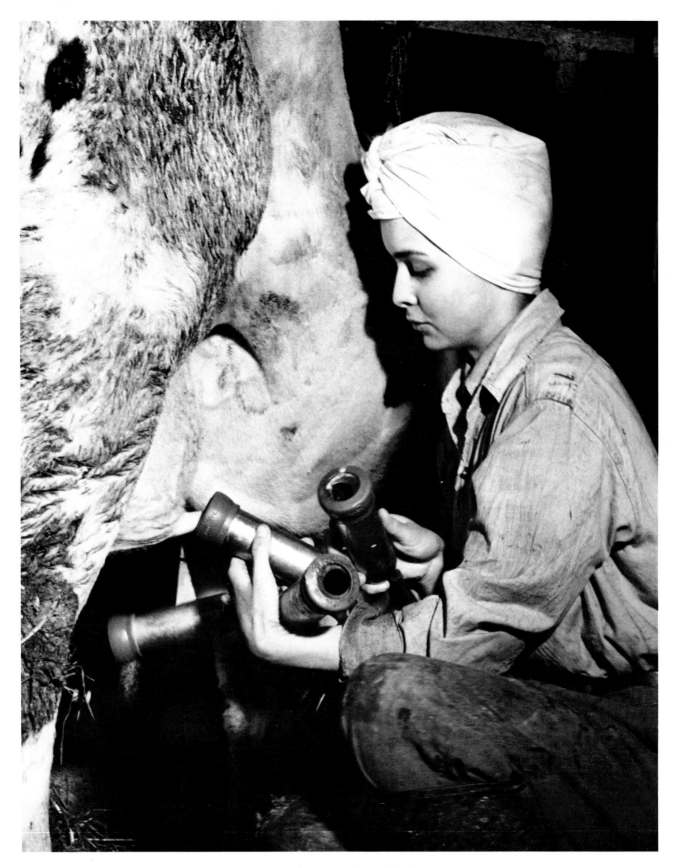

Because the giving of milk is a delicate physiological function, triggered by hormones, cows are treated gently. Most milkers know their cows by name and often speak softly to each one during mechanized milking, as does this New York State dairy farm daughter in 1948, removing the suction cups of the milking machine. Through better feeding and artificial insemination, milk production about doubled in this period, and by 1975 stood at an average of 10,354 lbs. of raw milk per cow per year.

Mrs. Len Cole, farm wife, Michigan, 1947.

Aerial dusting of insecticide on a potato field near Lodi, California, 1962. To farmers in the post-war period, pesticides were a boon that turned into a mixed blessing. Their increased use triggered government action because of contaminated water supplies, sickened farm workers, and the increased resistance of the very pests the chemicals were supposed to eradicate. Today research is moving toward biological control through natural predators, and biotechnology through the infusion of pest-resistant genes into plants.

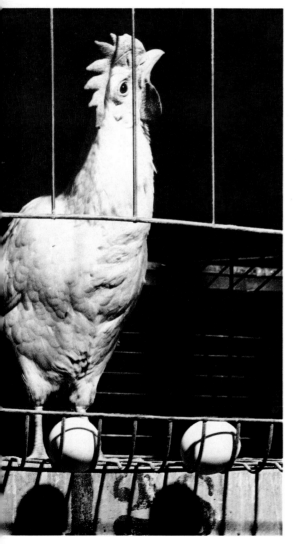

By the end of the 1960s there were poultry ''factory'' buildings where millions of broilers, and hens for eggs were raised in confinement in wire cages, with automated feeding and watering. The ''factory'' owners often were under contract for processors and distributors, an economic system known as vertical integration.

As World War II came to a close, many small farms still kept a flock of chickens for family eating, and the selling of a few extra eggs. Some small flocks had the run of the farmyard during the day, and were feed-trained to return to the henhouse nests for laying eggs. Flocks often tended to have a "boss" hen to lead the way around the farmyard looking for food, Ohio, 1950.

Farm cats and newly born lamb brought into kitchen during cold winter night, Ohio, 1948.

Bringing a boar to a mating session, Missouri, 1960. Geneticists and breeders remade the hog from a chubby, short lard animal to one that is long, heavily muscled, with large leaner pork chops and bigger hams.

Farmer brings flocks of geese to cotton field for weed and insect control near Bakersfield, California, 1965.

Milking goat awaiting herdsman, Ohio, 1946. Throughout the U.S., as a service to those people who are allergic to cow's milk, there is a small dairy business in goat's milk, butter and cheese.

A corn combine in Nebraska empties its hopper into a field wagon. In the mid-1940s the U.S. harvested about 3 billion bushels of corn annually; in 1975 when this photograph was taken it was about 6 billion bushels. Yet in the mid-to-late 1980s the harvest was only 6.5 billion bushels in the average weather years, an indication that the technological advances of the thirty years after World War II had begun to level off.

Spraying insecticide on a newly seeded lettuce field near Salinas, California, 1970.

Life on the Changing Farm

By Kirby Moulton

"One generation passes away, and another generation comes, but the earth abides forever"

Ecclesiastes 1:4

The changes in farming shown in Joe Munroe's photographs had roots in far earlier periods, but became strikingly evident between 1945 and 1975 as farmers used new technologies on a massive scale. At the end of World War II, almost all farms had electricity, workhorses were gone, tractor use was widespread, hybrid corn was being adopted and nitrogen fertilizers were poised for dramatic growth. Agriculture was truly a mix of the traditional and the modern. By 1975, farms were fewer but larger and required more capital. Rural and urban values had grown more alike.

Agriculture changed dramatically. Yields per acre for most crops increased with new genetics and improved cultural methods. Average wheat yields climbed from 17 bushels to 30 bushels per acre, corn production went from 38 to 87 bushels (160 bushels in some regions), and cotton output expanded from half a bale to one bale per acre. Milk production per cow more than doubled. This was a time of unparalleled growth in farm efficiency: a single farm worker in 1950 produced enough food to feed 15 people, but by 1975 produced enough for 50. During this period, the expansion in farm output exceeded the declines in inflation-corrected commodity and livestock prices, so that "real" farm revenues increased.

There were other changes as well: a loss of 4 million farms, an emigration of close to 16 million from the farm population, and a drop of over 7 million farm jobs during the three decades after the war. The decrease in farm labor went hand-in-hand with increased use of machinery and chemicals. Fertilizer and pesticide use per acre rose over 6 percent annually in the post-war years; Results often were inefficient because the chemicals were applied more rapidly than yields were growing. Machinery and chemical costs climbed faster than revenues, forcing farmers to expand their operations and seek ever higher levels of yields and efficiency.

These changes came at a tremendous human and capital cost. As people moved out, capital moved in—forcing up the value of an average farm by tenfold during the post-war period. Large capital investments were needed to buy more acres or more animals so that new, high capacity machinery could be used at its most efficient levels. These investments became stiff barriers for many farmers wanting to expand or seeking to enter farming for the first time. Those who couldn't afford or keep up with new technology had to leave agriculture. Those who remained became increasingly sophisticated business managers.

It was the number and rate of such changes that set apart the post-war period from earlier and later periods. If a bundle of statistics were combined for such factors as the use of chemicals and machinery, farm size and value, grain yields, and people fed per farm worker, and then plotted to a line on a graph, the line would rise quite consistently and sharply from 1945 to 1975. It would then drift into a sligthly rising but more leveled curve up to the present. A reverse curve would be apparent for farm numbers and employment which declined more rapidly before 1975.

As it turns out, this thirty-year period, during which Munroe was most active in farm location shooting, was the watershed in American agriculture, finally separating the practices of the past from the business agriculture of the future.

Over the past century, four major technological revolutions have affected agriculture, and their influences were particularly apparent during the thirty-year post-World War II period. The first was the mechanical revolution that replaced men with machines. The second was the chemical revolution that boomed yields per acre. The third influence was biotechnology, such as genetic engineering with crops and livestock, that increased disease resistance and productivity. The fourth influence was the communications and computer revolution, leading to better farm management.

Machines Begin to Take Over

Rice harvest, Colusa, California, 1961.

The mechanical revolution altered the way America's crops were planted, cultivated, harvested and processed for the market place. Machines allowed one man—with a lot of help from machinery manufacturers, oil companies, food processors, and transportation firms, among others—to produce enough food to feed over 80 people within the U.S. and other countries in the early 1990s and an expected one hundred people as the nation enters the next century. In the period between the two World Wars, new implements were designed, built, and refined.

During the three decades following World War II, however, the use of tractors and other mechanized equipment really proliferated. This adoption of mechanical technology had a profound effect on agricultural productivity, capital requirements, and the need for farm workers—as well as on soils, water and ecosystems.

An Iowa farmer describes life on the farm in 1946: "During the days, I'd work at planting, cultivating and harvesting. We cultivated corn five times a season to keep out weeds; we didn't have chemical herbicides. By 1946 we had a rubber-tired tractor, but still planted corn with horses pulling a two-row planter. To harvest we used a two-row picker mounted on the tractor; self-propelled combines hadn't appeared yet. For the tractor we had a six foot power-takeoff combine for oats, along with a hay mower, hay loader, rake and pitchforks. We had a pickup truck and one auto, a 1940 Chevy."

The change up to 1946 was slow compared to what came in the succeeding years. For example, the replacement value of farm machinery on the average U.S.

farm jumped eightfold in the following three decades—after correcting for inflation. Nationally, numbers of the most basic farm machine—the tractor—increased from 2.3 million in 1945 to 4.5 million in 1975, even though harvested acreage declined slightly. This expansion in tractor numbers was accompanied by increases in horsepower, four-wheel drive and specialized power take-off attachments, not to mention items such as shock absorbers and more comfortable air conditioned, radio-equipped cabs.

Huge clanking machines called "combines" became widespread largely for the harvesting of grains such as corn, wheat, soybeans, sorghum and rice. They combined the steps of cuttting the plant, stripping the grain from the ear or stalk, auguring the grain to a storage bin and blowing out the chaff behind the machine—all this in one action while moving through the fields.

California and some other areas became vast research laboratories for the design and production of machinery for working in cotton, sugar beets and a variety of fruits and vegetables. The cotton harvester was first used extensively in California, picking as much in an hour as a handpicker did in 72 hours. Mechanical harvesters were invented and swiftly adopted for carrots, potatoes and sugar beets. Large machines for lettuce harvesting and packing became common. These machines, with their 25 workers for cutting and packing the lettuce, replaced 50 workers in the fields and packing sheds. They cut and sorted the let-tuce and field-packed the heads in polystyrene film, ready for shipping.

The first grape harvester rolled into a California vineyard near Bakersfield in 1963. Tree shakers, ground sweepers and large, hooded vacuum gatherers became standard operating procedure in harvesting California's nut crop. In response to uncertainties about the continued availability of Mexican bracero workers for hand-harvesting chores, the tomato harvester was refined and adopted. It was made possible by earlier genetic work that developed strains of tough-skinned canning tomatoes which ripened all at the same time and thus were suitable for mechanical harvesting.

California was at the lead of mechanical innovation partly because its farming practices were not as deeply rooted as other regions in the past. Its climate favored the production of diverse speciality crops that, without machines, required relatively large amounts of labor. But the seasonal labor supply in California became uncertain during the post-war years as alternative job opportunities arose and regulations affecting the migrant labor force changed. The uncertainty of labor availability focused growers' attention on ways to get by with fewer workers. The farming environment was one of change and farmers were open to mechanical innovation.

By 1970, the two-row corn planters, commonly seen in the Corn Belt states in 1946, had given way to twelve and sixteen-row planters, many equipped to apply herbicides, insecticides and fertilizer at the same time the seed is planted. Larger and more efficient drills, discs, mowers, and sprayers were also perfected. Automatic cattle and hog feeding systems using augers and belts replaced the shovel and pitchfork. Mechanical systems capable of quick manure removal modernized the handling of solid animal wastes, though their increased capacity intensified the environmental impact of sizable fecal concentrations. Electronic sorting machines made fast and accurate grading and sizing of farm products possible without large work crews.

Balers, stackers, rollers, and other new equipment revolutionized hay-making systems. Machines now cut, bale, tie and stack hay in units ranging from the ordinary 50 to 100 pound rectangular cubes, to huge round bales 6 to 8 feet in diameter weighing up to a ton or more. Some farm operations cut the hay—usually alfalfa—and compress it into bite-sized cubes and wafers for direct feeding to cattle. By 1975, the value of the hay crop in the United States had risen to $6.5 billion.

Increased use of herbicides and pesticides created opportunities for aerial application. The early planes were Stearman biplanes that had been used as trainers during World War II. These surplus aircraft were fitted with spray rigs and adroitly maneuvered over fields, dropping quickly at one end to spray and climbing just as rapidly at the other. Later, specialized aircraft with greater capacities were developed. In the sixties, some growers began taking advantage of helicopters' powerful blade-induced downwash and turbulence to force chemicals into dense stands of crops.

The mechanical revolution led directly to more recent developments such as the use of laser beams to guide field leveling, to assure precision planting and to allow more accurate cultivation. The ability to cultivate corn, sugar beets, vegetables and other field crops with laser precision has become important as the use of chemical herbicides has been increasingly restricted.

Mechanization on the larger farms replaced some of the most unpleasant farm jobs, such as digging vegetables, potatoes and sugar beets at harvest. It also reduced the number of workers needed per acre or per ton of production. This shifted production and farm jobs to states that readily adopted new machines, and away from the states that did not.

During this period wages and working conditions improved for hired farm labor, but not without the conflicts that characterize periods of change. Farm labor/management disputes led to the rise of farm labor unions, initially in processing and shipping activities and organized mostly by the Teamsters. Later the activity spread to the production level, organized by leaders such as Cesar Chávez and others who used consumer boycotts in addition to strikes to achieve their objectives.

Results of labor organization were mixed, with the subject steeped in controversy. Although farm wages remained below those in the industrial sector, they rose from about 73¢ per hour in the late 1940s to $4.32 hourly in the late 1980s. This increase was slightly greater than inflation, but less than gains in industrial wages.

Increased farmer awareness and the actions of farm unions, operating in a favorable political atmosphere, stimulated better employment and living conditions in agriculture. Improvements were apparent in farm labor housing, sanitation, safety measures such as bilingual labels and stricter controls on the handling and application of agricultural chemicals, and the banning of the short handled hoe for weeding. Worker groups pushed for, and achieved, better coverage for farm workers by welfare programs and increased acceptance of their children in local schools.

Concerns about farm labor continued even after the loss of jobs to machines slowed. They focused more on working conditions and ways to improve the stability of the labor force. There was wide agreement that much should be done, but continued argument over how to do it.

Farming with Chemicals

Chemical spraying in Ohio, 1947.

Perhaps the most significant event in the chemical revolution was the development of a relatively cheap process for production of synthetic nitrogen fertilizers. This was soon followed by a steady flow of new chemical insecticides, herbicides, and other pesticides. The use of these chemicals transformed farming practices after World War II and improved the quality of crop products. It also, however, contributed to subsequent surpluses of grains, soybeans and cotton and to potential environmental damage and health problems for farmers and consumers.

Nitrogen has always been an important plant nutrient but until 1946 was not always available except at high cost or by plowing nitrogen-fixing legumes, such as soybeans or alfalfa, into the soil. During World War II the system of economically extracting nitrogen from the atmosphere was perfected. This set the stage for the rapid development of nitrogen-based fertilizers and their easy availability to American farmers. In 1950, farmers applied an average of 6 pounds of nitrogen per acre planted. By 1975, the average application rate had jumped over eightfold, to 52 pounds per acre. There were significant increases in the use of phosphate and potash as well. After 1975, usage rates tended to stabilize and even decline in some years. The extensive application of these fertilizers, coupled with the new plant varieties developed in genetic laboratories, led to a dramatic expansion of agricultural production in the U.S.

Herbicides, fungicides, insecticides and

other pesticides came into widespread use during the period and led to significant gains in farm productivity. Efficient herbicides permitted the effective use of "minimum tillage," a practice of shallow plowing that lessened the disturbance of soil surface and helped keep down erosion by reducing the need for deep plowing as a means of controlling weeds.

Large spray rigs became commonplace in orchards, working their way along the rows, often shrouded from view by the fog of their spray. In the Plains, the West, and especially California, aerial application became more widespread, with low-flying aircraft commonly seen winging across highways as they approached adjacent fields. Soil fumigation was needed for more expensive crops like strawberries and grapes to rid the soil of nematodes and other pests. Fumigation rigs spread long sheets of plastic film to cover fields and prevent evaporization of the fumigant. From above, these fields appeared to be coated with strips of shiny enamel. Before picking, cotton was defoliated chemically with carefully-timed air or ground applications. Cotton culture from the days of hand labor had changed forever.

The benefits derived from agricultural chemicals were accompanied by problems relating to increased pest resistance, diminished soil productivity, growing water pollution and toxic residues. Insects build up immunity to certain pesticides and sometimes require large dosages or new chemicals for their control. These chemicals, in turn, lose their effectiveness and must be replaced. Continued heavy use of nitrogen fertilizers may tend to harm the biotic elements in soils and lead to compaction, poor water penetration and related problems which ultimately diminish productivity.

The pollution of run-off waters caused by fertilizers and other agricultural chemicals may damage fisheries, drinking water supplies and down-stream users of irrigation water. Toxic residues pose dangers to farm workers (a problem recognized relatively early), and raise concerns about food safety (an issue of more recent origin). These problems are yet to be fully evaluated and were considered seriously by only a few people, such as biologist Rachel Carson, during the 50s, 60s and 70s.

Over time the issue of toxic accumulation, pest resistance, and food safety brought into question the wisdom of heavy chemical use. As side effects of agricultural chemicals became apparent, governmental regulatory measures increased. Squeezed by increasing costs and subject to more regulations, farmers began to use chemicals more carefully. Rather than adhere to an inflexible application rate and timing schedule, they adopted monitoring systems that adjusted quantities and timing to specific crop needs. In this way, farmers reversed the three-decade decline in crop productivity relative to chemical use (i.e., chemical use had been increasing more rapidly than crop production) and that resulted in more efficient chemical use after 1975. They couldn't have done this without the knowledge and experience gained in the 30-year period beginning in 1945. In time more farmers became interested in "sustainable" agricultural systems needing less chemicals and resulting in fewer undesirable residues, by-products and effluents.

New Plants - Better Cows

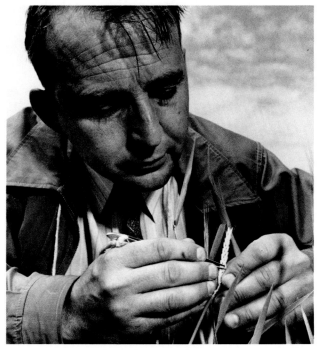

Crossbreeding barley, Kentucky, 1948.

The third revolution in post-war agriculture was biotechnology. While often thought of as a modern concept, biotechnology reaches back through history to the first attempts to understand plant breeding and develop new varieties of plants. The breeding of new varieties of corn, soybeans, wheat, rice, and diverse fruits and vegetables served to raise U.S. food and fiber production far beyond levels needed to meet domestic needs and set the stage for the United States to become a world supplier of food. Productivity gains were achieved in

the livestock sector, as well, and pushed up the output of American meat and dairy products. Work in genetic engineering, begun during the post-war period, holds promise for still greater production—along with the threat of human or environmental damage from inadvertent release of uncontrollable bio-engineered organisms.

The development of hybrid corn was a major force in post-war agriculture in America. Before hybrid seed, the customary harvest practice was for corn farmers to set aside a stock of ears in a crib, to be shelled and the kernels used as seed for the next crop. This incestuous system was haphazard at best, and did little to improve yields or quality.

Attention focused on corn because it was, and is, America's premiere crop. Corn is grown in every mainland state but much of its production is concentrated in Iowa, Illinois, Indiana and Nebraska. It occupies more acreage than any other crop, ranging between 60 and 70 million acres, and generates upwards of $15 billion in cash receipts for farmers in most years. Ninety nine percent of American corn is used for livestock feed—it never finds its way to the dinner table. The corn that we eat is called "sweet corn" and is grown specifically for our meal-time use.

The genetic manipulation of corn is physically much easier than with other grains because the male part, the pollen in the tassel at the top of the plant, is separated from the female part, the silks in the ear below, by a couple of feet. One grain of pollen falling on one silk equals one kernel of corn. It is a simple matter to cut off or redirect the pollen movement in order to cross-breed for hybridization with another variety. With most other grains the male and female parts are nestled together in the same flower. Out of thousands of in-breedings and crossings controlled by corn breeders each year, a few selected pure-bred strains are cross-bred in quantity. Plants from the resulting hybrid seed exhibit remarkable growth and vigor.

By 1975, well over 95 percent of the 6-million-bushel annual corn crop was from commercially produced hybrid seed—a bonanza for seed companies because the breeding must be re-done each year. Of course, this situation may change if current research into the hybridization of a perennial corn allows seed to be carried over for more than a year. Research results in the early 1990s indicate that success in this endeavor may be years away. Scientists, however, remain persistent in their search.

Other crops were by no means neglected. The breeding of new soybean, wheat, sorghum, and rice strains resulted in increased yields, better resistance to wind, and easier harvesting for the farmer.

Soybeans are a modern miracle crop, typical of the changing times of our land. They provide the high protein needed for food and feed, and nitrogen for revitalizing agricultural soil. In the 1930s two-thirds of the soybeans in the United States were plowed under to improve fertility by "fixing" more nitrogen into the soil. By 1975 the value of protein and oil for human and livestock food had become well recognized.

U.S. soybean production more than tripled in the three post-war decades; it stood at 2 billion bushels annually and had become America's number one agricultural export—a classic result of improved genetics and cultural practices.

Plant breeders also sought to increase the drought-tolerance of crops. It takes 646 pounds of water to produce a pound of soybeans, 575 pounds for a pound of wheat, and 348 pounds for a pound of corn. Plant geneticists are developing crop strains to resist drought (or to get along with less irrigation). Even as natural an enemy as dry weather has been mitigated by advances in biotechnology, especially in the West and other arid regions.

Many genetic improvements came at the expense of flavor characteristics; consumers claimed that "improved" fruit and vegetable varieties had a long shelf life but not enough flavor.

There were also worries about ending up with a narrow germ plasm base lacking the genes necessary to fight new pathogens. A germ plasm base is somewhat akin to a medicine cabinet. If it contains just the "miracle drugs", it might not have the remedies for new problems.

This concern led to increased concentration on the establishment and maintenance of diverse genetic stocks in seed storage facilities or other collections.

Antibiotics were developed which increased disease resistance for poultry and animals, allowing them to be raised in closed confined groups. This made feasible the massive poultry houses holding thousands of birds, the extensive cattle feedlots containing more than twenty thousand animals, and large-scale hog confinement facilities producing thousands of pigs for bacon and ham. All this led to questions about the safety—real or imagined—of antibiotic residues in meat and to public doubts about the humaneness of confinement practices in raising livestock.

Toward the end of the post-war period embryo transfers were being used by progressive cattlemen to create a pool of breeding stock. Hormone injections are used to increase the number of eggs in purebred cows. The eggs of these donor cows are fertilized and implanted in sturdy, but lesser quality cows, called recipients. This process can produce eight to fifteen calves per year from the high-quality donor cow instead of just one through natural breeding. This system keeps high-quality cows ovulating and conceiving more often and speeds the establishment of more productive herds.

Research into the digestive systems of farm animals, particularly ruminants, helped formulate improved feed rations using increased proportions of roughage from materials formerly discarded.

The application of biotechnology progressed rapidly in the 1980s and 1990s as ideas about the transfer of genes between species were tested in efforts to improve production, as in the case of bovine somatotropin (bst); or to permit the transfer of human genes to other animals (for example, cows) to allow the measurement of environmental impacts on those genes. These ideas have raised controversy, making their ultimate adoption uncertain.

While cow-calf operations have not changed dramatically over the years, the business of feeding beef cattle—once they are off the range—has undergone much technological change. Beef bought by today's consumer in the grocery store is seldom directly from the open range. Huge cattle feedlots fattening 25,000 or more yearling range cattle at one time, using computer-formulated rations, became a very visible (and odoriferous) evidence of the post-war agricultural revolution.

The Divining Rod to the Computer

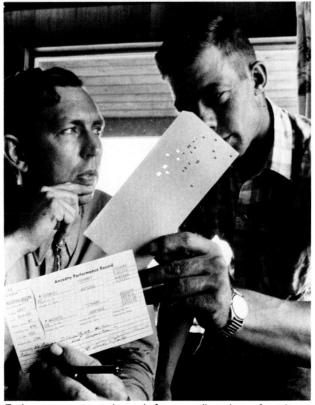

Early computer punch cards for recording data of cattle on the Codding Ranch, Oklahoma, 1960.

The fourth major influence affecting agriculture was the communications and computer revolution. This allowed for determination of optimal feeding mixes, more profitable combinations of crops, and accurate assessments of investment and marketing opportunities. This revolution led to better farm management and a more efficient agriculture.

Better telephone systems improved farm communications, as did television and radio-telephones. Farms progressed from a crank telephone and little electricity prior to 1938, to a party line and radio in 1946, to television in 1955, and finally, by 1975, to some computer use.

In particular, the advent of better communications led to a proliferation of market news reports. The reports were prepared primarily by government agencies, although private organizations were important in livestock and poultry reporting. These reports provided farmers with information such as nationwide crop planting, export-import figures and commodity prices and projections. Farmers used this information in making production and marketing decisions.

Early reporting services relied on the radio and mail to get information to farmers. During the latter part of the post-war period, communications used on-line computer service, recorded telephone messages, and television programs. Video tapes and satellite TV conferences were introduced to help farmers in using information and making decisions.

Computers became an important farm tool during the post-war years. They helped farmers analyze the information provided by improved communications. One of their early uses was to determine optimal feed rations for livestock, particularly dairy and feedlot beef herds. Another early use was to give analyses of animals' physical characteristics for dairy herd improvement programs.

California cattle feedlot with 25,000-head capacity uses automatic self-unloading feed trucks, 1965.

Computers and mechanical systems altered the nature of much hog farming in the United States. Hog raising has been a mainstay of U.S. farming, and just about every farmer has his own ideas about feed rations, housing, breeding, and farrowing techniques. During the three decades covered by this book, there were numerous innovations in large-capacity setups with several thousand swine being confined in buildings and small outdoor lots so that feed, water and manure disposal could be automated and controlled by computer.

Pigs on Ohio farm, 1951.

In addition to determining feed rations and crop mixes, computers were used to maintain detailed records of dairy production and provide information needed for financial statements and tax returns. The prodding of bankers probably explains some computer use on farms. Banks demanded more detailed analyses to support loans than in the past, and many farmers

bought computers to satisfy this need. As farmers recognized the help that computers gave in this task, they began applying them to other farm decisions. Despite the apparent advantages of computers, there are still farmers who say, perhaps with tongue in cheek, "computers are nice if you have time to wait for the high speed results."

These practices, even those adopted prior to computers, permitted farmers to gain a better idea of the economic consequences of various decisions. They could determine an optimal crop mix with new precision, figure out the benefits of conservation practices, and decide which technologies to adopt. As younger generations took over the management of farms in America, they made farm decisions using techniques learned from computer courses in their schools and universities.

The four major influences we've identified, plus other economic and social trends, resulted in a different face for American agriculture. In some ways there was a resemblance to the older face—because, after all, agriculture is made up of largely immutable, natural cycles of germinating, blossoming, pollinating, maturing, and harvesting. But in other ways there were new features, in that farms became fewer and larger, machines bigger and more sophisticated, and farm communities more consolidated and less self-contained.

All these developments didn't take place at the same time throughout the U.S., as Joe Munroe discovered while on assignment in Alaska in 1947. As World War II ended, homestead land in Alaska's Matanuska Valley cost $1.25 per acre. This secluded valley floor comprises 1,200 square miles and is located about 60 miles from Anchorage. It is ringed by the 7,000 foot Chugach mountains that serve as a shelter from the frigid winds from the north. An opening to the coast toward the southwest admits air from the warm Japanese ocean current along the Alaskan coast. This peculiar terrain and unusual climate made the Matanuska Valley and several similar temperate areas the only important farm regions in Alaska. Farms in this area concentrated on potatoes, hay and truck gardening. With the short growing season and long daylight hours in the summer, vegetables were sometimes abnormally large. Even so, Matanuska old-timers say that there are only two seasons: Winter and Fourth of July.

Alaskan farms were often rugged, with few amenities; like the early farms in the Great Plains. Clearing of large tracts was necessary to open up areas for planting. Sophisticated farming methods being adopted in the "lower 48" were hardly appropriate to the pioneering of new farms in Alaska. However, there was scientific progress, even in this remote region, as plant geneticists began breeding new strains of vegetables, legumes and grasses adapted to Alaska's soil and climate.

Even in America's primary agricultural regions, some aspects of farm work and rural life remained rooted in the past. A demanding work ethic still dominated the farmer's daily life. Easier transportation and mass television were bringing farm and city life closer together; even so, there continued to be feelings of neighborly dependency stemming from the comparative remoteness of farms and small towns. In the fields, gardens, orchards and vineyards, and in the day-to-day handling of livestock, one could sense that many of the sounds, sights and smells of farm life don't change. There is a continuity to the farmer's deep-seated respect for the land and his animals.

A Personal Look

It is fairly simple to assemble statistics about the changes in farming. But numbers alone often obscure recognition that these changes, no matter how broad, still affected individuals—people with differing hopes, fears and abilities to deal with change.

There are photographs and captions here from the length and breadth of the nation. Certainly no farmer or groups of farms could be called "typical." But out of the many dozens of interviews and research projects that Munroe did between 1945 and 1975, I've selected two for occasional text reference that seem to capture what was happening. One focuses on a mile of farm road in south-central Michigan and the other concerns the Raffety family of Iowa. In what follows, these experiences and observations are interwoven, along with broad-based nationwide information.

Changes in a Mile of Farms

One might say that this story began over 150 years ago in the midwestern Corn Belt states. In 1837, Isaac Fulton, a restless farmer in New York State, received a pur chase deed from President Martin Van Buren for 80 acres of farmland, sight unseen, in what was then known as the "wilderness" of fertile southern Michigan, along the northern edge of the Corn Belt. Along with many families migrating West, Fulton, with his wife on the seat beside him and a coop of chickens tied to the tailgate of his horse-drawn wagon, found his way over Indian and trapper trails to the surveyor's stakes marking his parcel of land near what is now East Lansing. Like thousands of others in America's heartland, he cut logs for the new homestead and began farming. The Indian trail was soon a rough road, and within a year or so along a mile of that road that included Fulton's place there were seven farmsteads, all pretty much alike. The sun and rain fell on all alike and the soil was the same quality on all seven. But as the years passed, when the harvest came there were full granaries and well-tended buildings on some farms and scrawny livestock and empty pantries on others. The communal spirit necessary to the pioneers was gradually giving way to individual enterprise, the pulling ahead of the more resourceful and energetic.

In 1947 Joe Munroe completed a research project for *Farm Quarterly* magazine along one mile of the farm road that included the old Isaac Fulton farm. Photographs from that project are scattered throughout this book. In 1947 there were still seven farms along that one mile. Some were good, some worn out; some big, some small. Munroe returned to this same stretch of rural road several times over the

years. Thirty or more years later there were still six farms. The Mile looked much the same as it did in 1947. But there were many changes beneath the relatively unchanging landscape—changes in farming techniques, land ownership and social patterns. What happened along the Mile of Farms has been mirrored throughout the landscape of American agriculture.

Seven independent farms operated on the Mile in 1947 and six of the owners lived there. By 1987, not a single farm remained as it was. Of the six farms still touching the

Alan Culham studies in farm house kitchen for master's degree in agriculture at Michigan State University, 1987.

Mile, four were part of larger operations with considerable land off the Mile and with owners living elsewhere. Two farms remained independent, one operated by part-time sheep raiser Alan Culham, and the other by retiree Dale Ball, who rented out most of his land. The average size of the operations touching the Mile, including land farmed elsewhere, had jumped to 1,500 acres in 1987 from 243 acres in 1947.

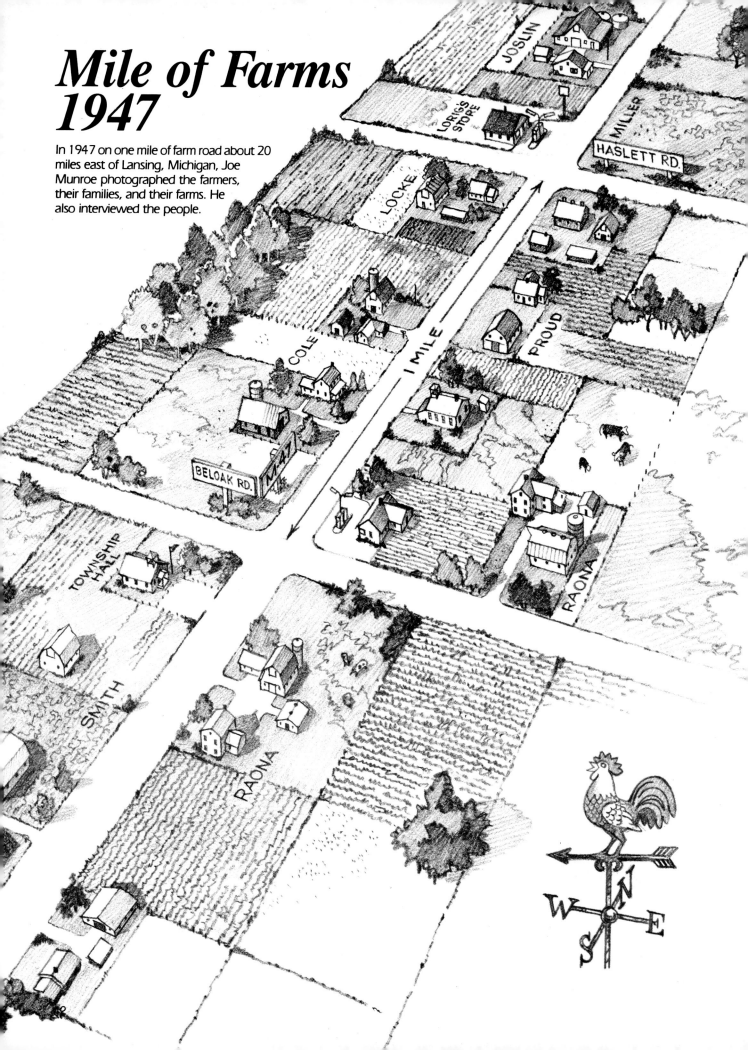

Mile of Farms 1947

In 1947 on one mile of farm road about 20 miles east of Lansing, Michigan, Joe Munroe photographed the farmers, their families, and their farms. He also interviewed the people.

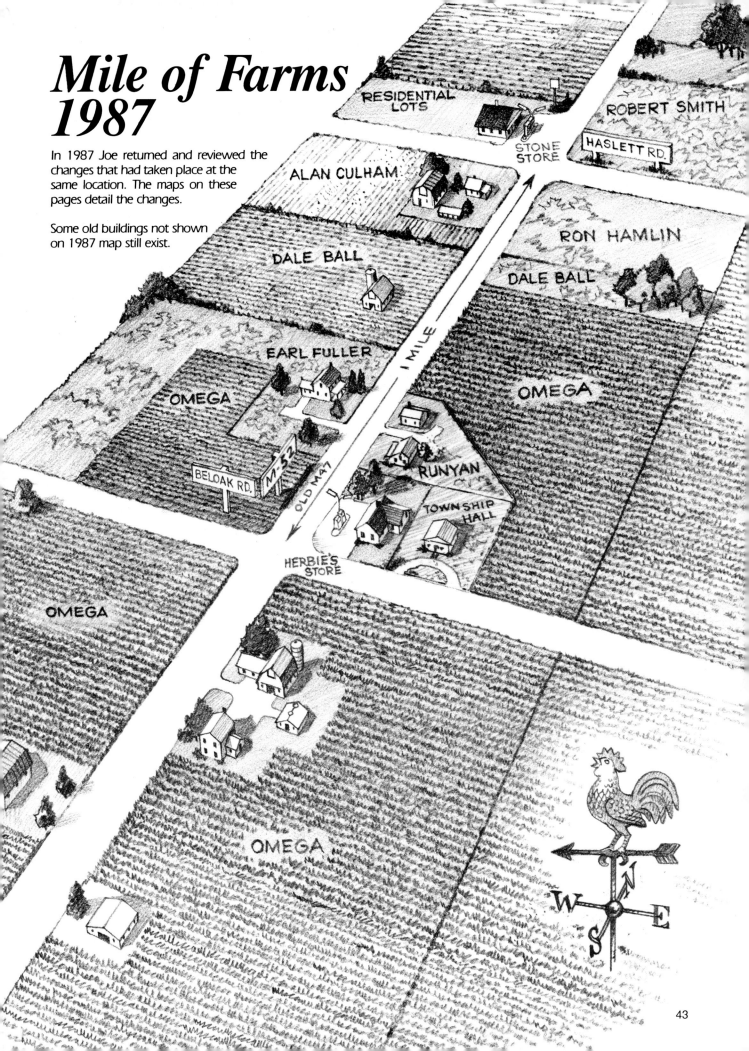

Mile of Farms 1987

In 1987 Joe returned and reviewed the changes that had taken place at the same location. The maps on these pages detail the changes.

Some old buildings not shown on 1987 map still exist.

The farmers and their several moneys, animals and acres on, or touching the intersections of...

The Mile of Farms – 1947

FARM OWNER	Acreage:		Gross Income per Acre	New Value Machinery	Persons Living on Farms on the Mile		Livestock
	On the Mile	Total on and off the Mile			Owners & Family	Others	
STANLEY SMITH. He was a philosopher.	235	235	$21.00	$ 6,075	4	0	12 dairy cows 50 chickens 200 sheep
DR. RALPH WADLEY, a city surgeon in East Lansing. RAY SMITH was his farm manager.	756	756	$56.00	$23,975	0	23 (Ray Smith and hired hands and families)	240 cattle mostly Angus 30 hogs 3 work horses
LEONARD COLE. Too old to do much work.	90 (60 acres rented out)	90	$17.00	$ 2,150	2	0	3 dairy cows 10 hogs 37 sheep
ORVAL PROUD. He was a hustler.	240	240	$37.50	$11,175	11 (Including hired men and family)		72 dairy cows 12 hogs
LAWRENCE LOCKE, part-time farmer. Worked in town.	118	118	$15.00	$ 2,800	4	0	7 dairy cows 3 hogs 250 chickens
JOHN MILLER, smart, and had two husky sons.	227	227	$97.00	$12,500 (Plus $4000 mint still)	6	0	8 dairy cows 35 sheep
DR. E.M. JOSLIN, a retired veterinarian.	40	40	Claimed farm had no income	$ 5,725	4	0	0
1947 TOTALS:	1706	1706	$34.71 Average	$64,400	31	23	

The Mile of Farms – 1987

OMEGA FARMS, owned by STEVE SIMMONS, an innovative executive. He bought the farms of STANLEY SMITH and DR. WADLEY and part of ORVAL PROUD's farm.	1,100	4,000	$227.00	$1 Million (For total operation)	0	7 (Hired men & families)	Up to 3,500 beef cattle mostly Angus, 6 working cattle horses
EARL FULLER bought what was once LEONARD COLE's place. Has a full-time field tile contracting business. Rents his farmland to ROY PIPER, a neighbor not on the Mile.	73	73	$ 15.00 (From rental)	0	3	0	0
DALE BALL, retired State Agricultural Director. He bought most of ORVAL PROUD's farm. Rents some crop land to ROY PIPER. Fishes in the farm pond.	160	160	$210.00 (Including farm rental)	$150,000	0	10 (Non-farm resident rentals)	50 beef cattle
ALAN CULHAM, studying for his Masters Degree. Bought what once was LAWRENCE LOCKE's farm.	57	117	$124.00	$ 30,000	2	0	Up to 250 sheep
RON HAMLIN, a hard-driving farmer, headquartered about 20 miles away. Bought parts of what once were ORVAL PROUD's, LAWRENCE LOCKE's and JOHN MILLER's farms.	98	2,900	$ 60.00 (From land on the Mile)	$750,000 (For total operation)	0	0	0
ROBERT SMITH, son of RAY SMITH, and the only farmer on the Mile today who lived here in 1947. Bought part of JOHN MILLER's farm.	40	1,690	$200.00	$650,000 (For total operation)	0	0	25 Angus beef cattle
1987 TOTALS:	1,528*	8,940	$139.33 Average	$2,580,000	5	17**	

Education: In 1947, 2 of the 7 landowners had education beyond high school. In 1987, 4 of the 6 landowners had some college education.

* Back portions totaling 178 acres of farms of Leonard Cole, Doc Joslin and John Miller have been sold during the 40 years and are no longer part of property that touches The Mile.

** About 5 non-farm, or "rural residents" live at the Runyan place, and possibly another couple or two live behind the Stone Store or next to Herbie's.

One Man's Farm Family

What happened on the Mile of Farms is what happened to U.S. farm land and how it was used over this changing period in American agriculture. What happened to individual farm families may be learned from the Laverne Raffety family, who since 1930, have owned, lived on and operated a family farm in the gently rolling land near Grinnell, Iowa, a few miles north of Interstate 80, in the heart of American's most fertile and productive agricultural region.

Laverne Raffety, Iowa, 1988.

Now retired, Laverne Raffety began farming in 1930 with his father's 160-acre farm, which at that time included 20 work horses. In 1946, at the beginning of the period covered by this book, the farm had grown to 320 acres. Laverne and his wife, Iness, raised two sons, Maynard and Howard, both of whom now actively operate the farm that, by 1975, was 1,200 acres. Munroe interviewed the Raffety family in Iowa, asking about their lives from 1945 to 1975, and many of their thoughts are quoted throughout the text. Several photographs of the Raffetys at work are interspersed throughout the picture collection.

The Raffety family faced the same decisions as farmers throughout the country: what crops will yield the best return? Should we enlarge? What about needed capital? Can we keep the kids on the farm? How they answered these questions is also how the mainstream of American agriculture answered.

The Results of Change

The three decades of technological innovation and implementation between 1945 and 1975 ushered in new farming practices, far greater productivity per farmer and larger-scale farming. These changes were supported by innovative farmers and rich land resources and permitted the United States to become a major food supplier to the rest of the world. But all of this was accomplished at the cost of fewer farms, a greatly diminished rural population, and a decline in rural communities.

Farmers responded in differing ways to the four revolutions that swept agriculture. Some chose to leave farming or to remain on a small-scale, part-time basis, others expanded to a truly large scale, but most chose a middle course of expansion that allowed them to maintain family control.

The four revolutions affected farmer decisions through economic pressures including high costs, low prices and intense competition. Beyond the farm, the revolutions affected the environment, rural communities, world trade and government policies. These changes pose dilemmas about how we would like agriculture to be in the future. Are large farms necessarily more efficient? What changes would be needed to move up to a more free-market farm economy? What policy changes might help maintain a family-farm economy? Are there ways to slow the agricultural treadmill that encourages expansion? What can be done for rural communities and to protect the environment?

A look at Michigan's Mile of Farms can help set the context for what is to come. The farmers on the Mile have been in the mainstream of change. They invest heavily in machinery and other farm inputs, they keep up with new technology, and operate as efficiently as possible. Today they raise fewer kinds of livestock and fewer varieties of crops than in 1947. They all have expanded in the cost-price squeeze and felt compelled to produce more in order to make a profit, or even stay in business. They all yearn for economic independence but find it essential to participate in government support programs. They welcome the growth of the international market but lament the degree to which this has involved agriculture in international politics. Like everyone connected with farming, they are simply trying as best as they can to live with the consequences of change.

How Farmers Responded

Some farmers were unprepared for the uncertainties of bigness, or of competing with their larger neighbors, so they sold out. This is clearly evident on the Mile of

Farms, where seven independent owner-operated farms in 1947 had been reduced to one by 1988. The others became parts of larger operations that were managed away from the Mile.

On a national scale, the number of American farms declined by almost 60 percent—from 6 million in 1945 to 2.5 million in 1975. The number of people living on these farms declined by almost 65 percent—from 25 million (or 18 percent of the total U.S. population) in 1945 to 9 million (or 4 percent of the total) in 1975. The total area devoted to farms, however, declined by only 12 percent—from 1.15 billion acres in 1945 to 1.01 billion acres in 1975. Nevertheless, average production of basic crops during the period increased by over 50%.

As many farmers moved out of farming, their lands were incorporated into larger holdings with little visible evidence of change except for the abandoned houses and barns scattered here and there. Howard Raffety noted that one-third to one-half

Howard Raffety, Iowa, 1988.

of the farmers in his area in Iowa left farming between 1946 and 1976. The emigration was of farmers, not farms. The land remained, consolidated into larger holdings like those of the Raffetys.

Other farmers, particularly in the poultry industry in the Southeast, opted to become units in vertically-integrated chains for large-scale poultry production. They essentially transformed themselves into factory foremen, responsible only for the production from their poultry houses, but with no say about management decisions. Except for poultry, with its relatively small land requirement, this system has been a small factor in the farm economy because it usually doesn't pay a processor or distributor to own an amount of farmland needed for other types of livestock. Some consider such consolidation of buyers as questionable because it reduces the competition of numbers of marketing outlets for farmers.

A few farmers opted for specialized large-scale production. In California, single vineyards of over 5,000 acres were established. Individuals and corporations farmed over 20,000 acres of specialty and field crops. Cow-calf and grass-fed beef operations had always been large because of the need for extensive range. But as management techniques improved and animal feeding and medication became more efficient, the focus of the cattle industry shifted to massive feedlots. Cattle were shipped long distances to be fed-out and slaughtered. Smaller feed lots disappeared and with them the smaller scale slaughter plants that they had supported. The loca-

tion of these lots and new slaughter plants were not dictated just by transportation and labor costs. Heightened environmental concerns forced the location of such facilities away from urban and suburban areas and significant watersheds.

Family operations, like the Raffety's in central Iowa, sometimes opted for a middle course. They increased their farming operations from 320 acres in 1946 to a little over 1,200 acres in 3 almost-equal principal parcels in 1975. The single family management had changed to a corporate structure owned by the father and two sons. This structure became more and more typical of American farming and certainly was not the intended target of those berating "corporate agriculture" in America.

Ron Hamlin fills herbicide sprayer, 1987.

Ron Hamlin of the Mile of Farms provides an example of entrepreneurial drive leading to larger farm sizes. Hamlin started in the 1950s with 40 acres and a contract hauling business. By 1988 he was operating 2,900 acres devoted to corn, soybeans, wheat and alfalfa. Included in this was 98 acres on the Mile of Farms that was purchased in 1983 as part of his expansion program and which consolidated farming operations formerly operated by two other farmers on the Mile.

The Economic Effects of Change

New technologies have tended to foster growth in the size of American farms. As these technologies increase production from each unit of land, they also help to depress the prices for that production. This in turn stimulates the drive for additional production to spread capital costs and thus lower unit production costs. Willard Cochrane, a perceptive ag economist from the University of Minnesota, noted that farmers ended up on what he termed the "agricultural treadmill." They had to run harder just to stay even. Robert Smith, the only farmer on today's Mile of Farms who was there in 1947, translated the treadmill into personal terms when he observed that today he has to plant 800 acres of corn to raise enough money just to pay bills.

Even though their returns on investment declined and their costs increased, those farmers who continued to farm made more money toward the end of the post-war period. The income left to America's farmers after paying expenses and after accounting

Robert Smith, Mile of Farms, Michigan, 1987.

for losses in inflation-adjusted purchasing power, had declined about 45 percent—from $94.7 billion in 1945 to $52.0 billion in 1975 (these numbers are expressed in 1987 dollars). But since there were far fewer farms in 1975 than in 1945, net income per farm, corrected for inflation, had increased by about 30 percent—from $15,865 in 1945 to $20,625 in 1975. The value of farm assets increased even faster, causing the rate of return on farm investments to drop from 12 percent in 1945 to 4 percent in 1975.

Farmers are competing against each other, selling to relatively few buyers that, in turn, have the comfort of a stable demand for farm products. Under these conditions, farm prices are pushed close to production costs and any economic advantages that farmers gain through new technology are transitory and rapidly given up to the buyers. Farmers need to continuously adopt new practices and technologies if they are to stay in agriculture.

The Environment and Sustainable Agriculture

Changes in agricultural practices affected the environment both positively and negatively during the period. Much of the farmland which seriously eroded during the pre-war Dust Bowl years was more carefully protected in the post-war period. But as "fence row to fence row" farming and chemical use accelerated, the soil erosion problem re-emerged, as did new concerns about water quality, worker safety and food standards.

The fight against soil erosion used contour plowing, terracing and crop rotation. Contour plowing and the terracing of hillside fields impede storm waters from churning down cultivated slopes and carrying away fertile topsoil. The rotation of crop plantings from year-to-year on any given field helps diminish erosion on sloping land when alternate-year use is in hay or pasture rather than in crops that require deep plowing. Crop rotation also naturally helps to control insects and diseases that would otherwise build an affinity for any one crop planted year after year. It reduces the need for synthetic fertilizers when the alternate-year crop is a clover or legume that fixes nitrogen in the soil and will be plowed under the following year. To stimulate protective practices, the government introduced the Conservation Reserve Program, which pays farmers for removing highly erodible lands from production for a 10-year period (a modern variant of the land bank concept).

Concerns about the environment led to an examination of "sustainable" agriculture—a set of agricultural practices compatible with very long-term agricultural productivity. In considering sustainable agriculture, it is well worth reflecting on the long history of the Mile of Farms. There are seven farms along that mile that have been dependent on the same soil and water resource for over 150 years. Drains were installed and crops rotated to protect that resource, just as the Raffety family plowed the contours to preserve their soils from erosion. Such practices need to continue to combat erosion and ease the high dependency on chemicals that may ultimately wear down soils and pollute waters.

Changes in Rural Communities

The forces of technology changed the practice of agriculture and with it the character of farm and rural life. As farms became larger, the houses became more like their urban counterparts. Farm chores became easier and life styles more modern. While many traditional values remained woven into the fabric of rural life, newer threads of thought were also inserted—just as in other nonfarm sectors.

The sense of community seemed to wither away. The supportive family and neighborly activities in 1946 described by Laverne Raffety became a way of the past. "Our neighborhood was close-knit. There was no television. We'd get together with neighbors to exchange help at silo-filling time," he remembers. "The women would put up preserves and do the cooking; the main meal was at midday. The big threshing dinner get-togethers had begun phasing out about 1946 as the machines began taking over. Sometimes there'd be Grange meetings in the winter with the men in the living room, women in the kitchen and youngsters in the parlor."

As roads and communications improved, fewer rural communities were needed to provide commercial support to farms. Increasingly farmers traveled to larger urban centers for their business and household needs. Many communities that had depended on farmers began to disappear or to become bedroom communities or light industrial centers. Visual evidence of this is clear in parts of California's long Central Valley as well as along the struggling main streets of Iowa and other Midwestern and Great Plains states. The decline of these rural communities has meant fewer opportunities for individual businessmen, and less community feeling to sustain the deteriorating areas.

Effects on Trade and Policy

The surge in U.S. agricultural production after World War II provided more food and fiber than could be used even in America's growing economy. In some cases, this excess production pushed the

United States into the center of global competition. In other cases, particularly in specialty crops, farmers made explicit decisions to serve foreign markets. Wheat, cotton, corn, soybeans, tobacco and a myriad of other commodities found markets overseas. This was good, but it increased the vulnerability of U.S. farmers to foreign economic and political changes. Agricultural exports grew from $3 billion in 1950 to $22 billion in 1975, with the U.S. ac-

Bulk rice storage, California, 1961.

counting for half of the world's grain trade. U.S. agriculture became a big business and farmers changed to accommodate this.

Expanded production also had its impact on government payments to farmers. Although cash receipts from farm marketings increased 11 percent, after correcting for inflation, between 1945/49 and 1970/74, direct government payments to farmers ballooned by 155 percent. Government price support programs that were designed to pull farmers from the depth of depression were heavily stressed to cope with the surplus problems of later years.

Furthermore, government-supported prices and domestic and foreign trade policies affected export markets. The U.S. embargo on soybean exports in the early 1970s, for example, forced customers to look elsewhere for supplies and contributed to the development of Brazil's soybean industry. For the short term, the effects of political tensions leading to embargoes, quotas, or escalated tariffs can cause price dislocations. Over the long term they lead to different production and trade patterns. American farmers cannot escape from the consequences of today's global food and fiber economy.

Some disagree about the long-term effects of political actions. Lauren Soth, editorial writer for the *Des Moines Register,* believes that political forces don't very much affect overall supply and demand. He claims that the U.S. embargo on grain shipments to Russia wasn't the culprit in the farm decline of the early 1980s. His theory is that figures show we exported as much or more during the embargo. His point is that, for example, we'd officially sell to Argentina; they'd change the bill of lading on the high seas and the grain would end up in Russia anyway.

The Dilemmas of Change

The changes to farming and its physical, economic and social environments involved some important personal trade-offs. Those people who wanted to remain in small-scale agriculture, or who could not afford to expand, found that they could not support themselves with only their farm income. Increasingly they had to find jobs away from the farms to supplement that income. Even in 1975, most American farm families received more than one-half of their income from off-farm sources. For the first time, money earned by U.S. farm families from off-farm sources was surpassing the level of that earned from all of their on-farm operations.

The changes of the past are viewed as positive by many who believe efficiency, low food costs, and better rural living standards satisfy society's value system. Others view the changes as negative because larger farms, declining rural communities, and widespread chemical use don't fit within

Sugar beet harvesting, Mile of Farms, 1947.

their concept of social values. The dilemma of values is expressed by the farmers of this book: the drive for efficiency by Iowa's Howard and Maynard Raffety, while still longing for the simplicity of their father's time; the desire to walk away from government programs as expressed today by Robert Smith of Michigan's Mile of Farms, and the fear that doing so would ultimately force farmers to "punch the clock for Safeway" as expressed by Maynard Raffety; and the wish, expressed by Dale Ball, to keep farms small and viable, perhaps as they were on the Mile of Farms in 1947; faced against the profit-driven reality of growth, described by Steve Simmons, that had changed the Mile by 1987.

There is a growing public concern over the diminishing number of small to medium full-time family farms. Is the trend to bigness inevitable? Will it endanger our ample, low-priced food supply or threaten environmental integrity or rural life values? Should we regulate farm size or direct tax money in the direction of the family farm rather than to all farmers large and small? These are social questions the people of the U.S. need to answer.

The argument about values seems to center around an agriculture of efficiency and free markets or an agriculture of farm families and small farms. And common to both visions of the future is a desire for a system that will stop the treadmill that has sucked the economic returns from farm investments, and the hope that it will do so while keeping agriculture as a stewardship in harmony with the environment.

Size and Efficiency

The continued rapid adoption of new technology is essential in a free market agriculture to keep ahead of lower cost competitors or those receiving government subsidies. But this tends to favor larger operations with the financial resources and diversification that permit early adoption of new technologies. Thus, a key strategy for surviving in a free market favors bigness. Just how big is open to question.

Many farmers seek technical efficiency and can operate on a large but manageable scale; a size that is far short of the truly big American farms. Ron Hamlin who farms 2,900 acres near Michigan's Mile of Farms, is an example of this. So is Iowa's Raffety family with their 1,200-acre operation. Steve Simmons is owner of the 4,000-acre Omega Farms that incorporates the old Orval Proud place on the Mile of Farms. He noted that he pays more just to fuel his machinery today than the cost of all the machinery on Proud's old farm in 1947. He commented, "It's unfortunate, but farms like those along the Mile 40 years ago just can't survive. You can't gross enough money out of 150 to 200 acres to own the machinery to do it."

Steve Simmons (center), Mile of Farms, Michigan, 1987.

Some farmers, while believing in large-scale techniques fostered by computer-based management, still adhered to traditional methods from the past. For example, Ralph Howe, farming in Iowa in 1968, was able to raise 1,100 hogs and 100 head of cattle on 320 acres of land using only fam-

Orval Proud (center) and Vance Mead, hired hand, 1947.

Similar changes occurred in dairy farming. Average dairy size increased from 5 to 6 cows just after the War to about 25 in 1975, as farmers sought to keep their revenues in line with increasing costs. Typical of this pattern, most farms along the Mile had a few milking cows in 1947 and one, the Orval Proud farm, had 72, then considered a fairly large herd. Forty years later, the closest dairy to the Mile was some distance to the north with a herd of 400 milkers, using not much more labor than had Proud. Dairy farming disappeared from the Mile of Farms just as it did from many once diversified farms throughout the country. It shifted to locations with better transportation links to urban areas, or with land that was more efficiently used for pasture and hay rather than grain.

Dale Ball affixes ear tag, 1987.

ily labor and an occasional hired hand. The farm grew all the needed corn, oats, hay and pasture. Two boars, purchased once a year, were the only hogs bought for the farm. Clearly, efficiency in 1968 did not depend entirely on the large scale of farm operations but it did require good management. This is still true for many of today's farms.

Figures show that, all things considered, giant farms are not necessarily more efficient than small or middle-sized farms, nor do they always have the same incentives for environmental protection. In 1985, for example, returns on investment for farms averaging 9,500 acres in size were sharply lower than for farms of less than 2,000 acres. Dale Ball, former Director of Agriculture for the State of Michigan and owner of 160 acres on the Mile of Farms in 1988, expressed the view of many farmers: "I hope we've learned something, and people will have a farm they can run efficiently and not become so big that they end up working for the machinery companies and the lending institutions and don't make enough to send their kids to college. I don't think you have to be that big to be efficient. In fact, I think there is a point of diminishing returns."

The Politics of a Free Market

Agriculture throughout the world is subsidized. New Zealand is an exception since subsidies were dropped a few years ago. Farmers there are still wondering how to compete under a new set of economic rules. A similar problem would exist here if subsidies were dropped to accommodate a free-market oriented agriculture. Yet there is no question that subsidies distort competition.

Maynard Raffety expressed the view of many in summarizing the need for farm supports. "Farming is full of unpredictable vagaries such as weather and diseases. We farmers love to talk about getting the government off our backs—to be free of production controls, but in most years we'd be in chaos if that were to happen. The U.S. has always had the capacity to overproduce," he asserts. "Yet there are times like the drought year of 1988 when we were probably fortunate to have a stored car-

Maynard and Eloise Raffety, 1988.

ryover from previous high-production years. Unlike, let's say, TVs or luxury automobiles, the nation's food supply is a basic necessity—not an option." If one studies history, in his opinion, "It seems apparent that laissez-faire agriculture doesn't work; it probably requires some sort of controlled niche in the free-market system. If the government were completely out of agriculture," Raffety concludes, "every farmer would plant all he could and he'd eventually be punching a time clock for Safeway and the meat packers."

There are two sides to the argument about getting the government off the back of agriculture. On the one hand, the government is the only impartial institution with the size and geographic scope to have a chance of smoothing out the production cycles by using such programs as the ever-normal granary, set-aside programs and subsidies. The government has walked the tightrope in seeing that farmers get decent prices with a reasonable price to consumers. It is different from farm protection programs in the European Community or Japan where consumers pay for farm supports through significantly higher food prices. In Japan, for example, the consumer price for rice is ten times what Americans would consider normal. By comparison, U.S. programs leading to inexpensive food for consumers have been relatively cheap for the taxpayers.

On the other hand, government programs result in a large bureaucracy with attendant red tape and political maneuvering. Sometimes, they are not an efficient way to disburse money. Getting paid for not producing is a form of welfare, but it seems necessary for the survival of a large class of U.S. farmers. Consequently, farmers will continue to accept the benefits of these programs, although with muted pleas to "get the government off our backs."

Steve Simmons (owner of Omega Farms, on the Mile of Farms) argues that while reducing government intervention in agriculture might be desirable, it presents a dilemma for everyone concerned. "If the government were to cease paying subsidies, agriculture would be thrown into turmoil for four to five years," Simmons contends. "If they had done it twenty or twenty-five years ago, we would have been squared away now, with no government subsidies, just supply and demand. But after 50 years of government grain programs, there's no way we can just all of a sudden drop them."

Roy Piper, another successful operator who farms for two property owners on the

Roy Piper at corn planting time, 1987.

Mile of Farms, would prefer to get along without farm programs but acknowledges that he cannot since the programs set the price of corn for him.

Improved technology raised his yields but escalated his costs as he bought larger and more sophisticated equipment.

"People paid their taxes to support government programs and universities to teach me how to produce better," he said. "Now, as surpluses develop, they're paying for price supports and crop storage because they taught me so well."

One way to break the bonds of regulation would be to adopt a program that divorced income supports from price supports. This is the so called "decoupling" scheme that would seek to solve the problem of low farm income through supplemental payments directly to those farmers with low incomes. Farm prices would not be touched under a decoupling scheme and would change as supply and demand conditions changed. Acreage restrictions would be dropped and farmers would base their planting and marketing decisions on their assessment of prices and other information. Under a decoupling scheme, farmers that are now supported under government price programs would make their decision just like the thousands of American farmers that are not supported. Regardless of how attractive such proposals may seem, most American farmers currently benefitting from government price support programs would have difficulty in adjusting to a subsidy-free system.

Today's free market system does not mean the elimination of all rules. Regulations affecting health and safety would continue. Improved market news services would be required to provide information about weather, crop and livestock prices, market movement, plantings and crop yields. International trade rules are needed to guide agricultural competition between different political systems. Rules protecting the environment and preserving natural resources would be essential.

The net effect of a free market agriculture in the coming decades might be similar to the changes within the period covered by this book: fewer and more efficient farms supplying the greatest part of the American food supply; better living standards for those farmers remaining; and low-cost food for American consumers. But there might also be a further weakening of rural communities; fewer farm families; and a potential loss of the farmer's philosophy about the "stewardship" of land.

The Politics of the Family Farm

The barriers against maintaining a family farm structure include the high costs of new equipment, escalating land costs, and the exodus of farm children to better paying and more secure jobs. Policy changes might help overcome some of these barriers. Perhaps there should be limitations on farm size, as there is in Norway; or perhaps farmland should not be taxed, a policy followed in England; or perhaps new technology should be ignored, as the Luddites of 19th century England proposed. A fundamental change in American economic policy would be essential in determining an acceptable farm size and enforcing it. Given the reluctance of government to control the size of automobile, steel or oil companies, it would be hard to imagine such a policy for agriculture.

Maynard Raffety analyzed the dilemma of discriminating in subsidies between "large" farms and "family" farms. "I do think there is a question of whether the same proportion of general taxpayer subsidy support is justified for the huge farms—especially in the West where the irrigation water supply is already a form of public subsidy," he said.

A pragmatic approach was suggested by Maynard's brother, Howard: deny the benefits of farm price supports to large-

Laverne and Michael Raffety, Iowa, 1975.

scale farms. This has been tried. The government limits price support payments to $50,000 per farmer, but these limits have been evaded by reorganizing farm ownership. Farms that are owned by a husband and wife can receive $100,000 in payments; if the farm includes children in the ownership, the limit equals $50,000 for each of the listed owners. Complicated partnerships are used to raise the payment limitation even further. This system has created a financial bonanza for lawyers and accountants and has not changed the fact that the biggest farms still get the most. The lesson is that a different policy with the specific

objective of targeting benefits to family-sized farms would be needed to alter the trend away from family and small farms.

Quality education has been tremendously important to the survival of farm families. It has provided up-to-date training in modern agriculture for those remaining on the farm, and it has provided alternatives for those leaving our rural areas. As Laverne Raffety observed, "Give a child land, he can lose it; give a child money, he can lose it; but give a child an education and he'll always have it." A system of family based agriculture in the United States would require a continued political commitment to quality education. The changes of the Mile of Farms indicates the important role assumed by education over the post-war period. In 1947, none of the farmers on the Mile had agricultural training at the college or university level. By the end of the period, four of the six farmers had at least college training.

Other policy approaches might include financial assistance to first-time farmers, a system of licensing to restrict entry into farming, or a control on farmland prices. Howard Raffety presented an argument for agriculture's public tax support by non-farmers. "Americans work 15 percent of their time to pay for their food; the rest of their time can be spent working for better living standards. People who have to worry constantly about where their next meal is coming from don't make very good workers, executives, lawyers, scientists or anything else. Farmers want to live well like anyone else. It sure seems to me that the average American has a stake in protecting a stable and healthy agriculture."

The agriculture treadmill will not stop of its own accord, and it is doubtful that any government policy can compel it to stop. As long as a large number of farmers sell the same commodity to a relatively few buyers, economic power will lie with the buyer. Our one million grain farmers sell most of their grain to five major private grain trading companies, including Cargill, Pillsbury and Continental. There are now only three very large private meat packing companies that virtually set the price for all beef and pork sold by U.S. farmers.

Slowing the Treadmill

There are strategies that might permit some farmers to get off the treadmill or to slow its speed. One strategy is to make a product different from others in the commodity class. Sellers of branded products, such as Del Monte or Foster's Farms, have succeeded in separating their products from other canned vegetables or fresh chickens. This usually results in premium prices for the product. The benefit for farm producers depends on their bargaining position relative to the brand owner. Some rice growers have acquired rice milling facilities and specialized in products like rice cakes or brown rice as a means for sidestepping the intense competition in the commodity market.

Another strategy is to form marketing cooperatives or bargaining associations that are large enough to deal with the largest buyers of agricultural commodities.

Land O' Lakes, the large dairy cooperative, is an example of how producers can gain market power through this strategy. The strategy seems to work in cotton and rice, where large cooperatives such as Calcot and Riceland Foods have market power, but is enormously difficult in other globally traded basic commodities like corn and soybeans. Marketing cooperatives are powerful also in almonds, oranges and processed fruits. Bargaining cooperatives are important in processing tomatoes and canned peaches and pears. The establishment of effective marketing cooperatives or bargaining associations is difficult for most American grain and livestock farmers partly because of their geographic spread and diversity, the enormous number of farmers involved, and the availability of large foreign supplies.

A restriction on entry to agriculture might tend to keep the number of agricultural competitors under control but would be unlikely to redress the balance between buyers and sellers. This remains an academic type of proposal because there is no chance that such a policy could emerge from the American political process.

The struggle against the treadmill is likely to remain as a significant characteristic of American agriculture regardless of whether a free market or a more "family farm" system of agriculture emerges. Why do farmers continue to put up with this discouraging economic process? Perhaps it is an abiding faith in farming that keeps some of them going; a faith expressed so simply by Laverne Raffety. "Farming has been a good life for me. It was good because there was work for all in the family, and all worked together. I've always wanted to pass on the farm for future generations better than it was when I took over. We've worked hard, but we've been lucky; the Lord smiled on us,"

Communities and the Environment

The future of America's rural communities will not be dictated solely by what happens to the number and size of the nation's farms. Many communities have become significantly less dependent on agricultural activities in recent years as services, manufacturing and tourism have increased in importance. If agriculture continues on the course of the past with fewer and larger farms, then a continued decline would be expected in those communities that rely heavily on agricultural activities. If agricultural trends or public policies change so that the current family-farm component is maintained or enlarged, then the viability of dependent rural communities would be enhanced. The communities would contain more owner/operators and fewer farm workers than they might otherwise have. The demand for smaller lots of farm supplies, financial services, and family amenities would increase and, potentially, support a more robust economy. Public policy changes would be needed in addition to farm changes to preserve communities. New policies could be targeted directly on problems of poverty, public services, and alternative economic opportunities.

Environmental protection became a tremendous public concern toward the end of the post-war period. Much of how agriculture responds to this concern depends on the awareness, attitudes and practices of America's farmers.

Good farming conveys a sense of man's respect for the land; almost a reverence for the great American resource. The choice between different farming systems doesn't imply a loss of this respect. An agriculture in harmony with the environment is a necessity for both the free-market and family-farm systems.

Maintaining agriculture's productivity requires the sacrifice of short-term economic benefits that can be gained from the rapid depletion of soil and water resources. Public policies will be essential in protecting those farmers adopting environmentally sound practices from the competition of those who do not. Policies must also assure that competitive pressures are not allowed to foster practices that damage the environment or endanger human life.

It's clear that agriculture will continue to change. It will be shaped by the dictates of economic necessity, formed by farmers' evolving values, and nudged and prodded by government regulations and social pressures. Pockets of tradition will remain, like the cow-calf operations of the Western range. And the appearance of the land and of its people, as captured in this book, will stay with us as a visual reminder that the American farmer may "own" the land today but in the "big scene" of events he is its short-term caretaker for the future. ■

A sweaty worker in California uses baby powder during a break in the rice harvest to ease the itch from chaff swirling from the huge machines. In 1975 the value of the U.S. rice harvest was $1.1 billion.

Work

Rice harvesting in the 1960s still was rough, hot work. Yet technology was changing the work patterns across the entire fabric of farm life. As one example, most of the nation's rice crop was being planted, fertilized and treated for weeds—all by air.

Corn kernels in farmer's hand, Illinois, 1958. With a unique combination of soils and climate, U.S. farmers devote more land to corn than any other crop, growing half the corn produced in the world. Genetic manipulation has hybridized corn, leading to large increases in yields as well as uniform plant sizes in commercial corn that makes it easier to harvest with mechanical combines. By the mid-1950s, hybrid rather than open-pollinated corn made up about 95 percent of all U.S. planting.

Scooped-out cross section showing corn seedling one week after planting, about to break through soil surface, Iowa, 1961.

Farm mechanization greatly increased in the thirty years after World War II. By 1950, most corn was planted by tractor-drawn two- or four-row mechanical planters, although some small fields were still planted by primitive hand-operated devices such as this one in California back country, 1958.

In 1975 in the Corn Belt about 77 million acres of corn were planted in a matter of days, using large units such as this twelve-row planter in Iowa. By this time, some planters were rigged to apply herbicides, insecticides, and fertilizer simultaneously while planting the seed.

Backyard vegetable gardens were once an important farm family food source but
have diminished in importance as roads, and easy transportation have brought farms
closer to supermarkets, and shopping malls (left), Kentucky, 1947 (right) near
Oneonta, New York, 1950.

An example of mixing large scale techniques with traditional methods. Ralph Howe, in 1968, with only occasional one-man hired help, raised 1,100 hogs and 100 cattle on his farm of 320 acres near Clemons, Iowa, with annual gross sales of $75,000. The farm grew all the needed feed of corn, oats, hay, and pasture. Two boars, purchased once a year, were the only hogs bought for the farm.

*W*hile sophisticated technology and labor-saving machinery are always welcome, it seems there are times when only human contact will do. Have crops ripened enough for harvest—too dry or too wet? Is this soil the right type? Are there signs of insects, disease? Is the animal sick? Has the feed mix blended okay? Farmers always have had an affinity for their land and their animals, and an everyday need to touch, inspect, smell, or even taste whatever they are handling.

An Ohio farmer in 1948 sniffs manure compost in which he grew commercial mushrooms in his moist, steamy cellars. Smell determined proper growing conditions.

Maynard Raffety, Grinnell, Iowa, 1975, walks into the soybean field and bites a kernel from a pod to determine whether the crop is dry enough to harvest without danger of moisture spoilage while in storage.

Another Ohio farmer in 1948 digs up a sod clod for inspection of the root system progress in his birdsfoot trefoil pasture.

A Michigan farmer in 1946 crumbles ripening barley heads and blows away the chaff, leaving the grain kernels in his palm for inspection.

In the post-World War II period, many descendants of Civil War era miners who had emigrated from Finland farmed the stubborn soil of Michigan's Northern Peninsula. It was a rigorous life for farming—sometimes only sixty days from frost to frost. Potatoes were the major crop but farm produce also consisted of dairy products, hay, and some wood from the farm woodlots. Finnish women worked in the fields with the men, as shown in these photographs of haying and firewood cutting near Marquette, Michigan in 1946; primitive methods when compared to developing mechanization.

As with any business, record keeping is a necessity on the farm. Computers would assume greater importance during the post-war decades, but on this purebred Correidale sheep farm in central Ohio in 1946 a handwritten ledger with pasted-in snapshots kept the data used to establish individual animal records for future sales.

In 1975 Maynard Raffety on his farm near Grinnell, Iowa, had a small desk calculator to help keep records of his dairy herd production. His wife, Eloise, in background, made homemade ice cream for dinner using an old-fashioned freezer.

A Montana rancher in 1964 noting progress of Hereford calf weight.

In the summer of 1966, thirsty sheep crowded around this Idaho sheep raiser as he filled the watering trough on the remote range from a flexible pipe leading from a tank on the pickup truck.

Sheep convert what would be wasted roughage—such as fields of grain stubble and sparse pasture—into wool and meat. The number of sheep and lambs in the U.S. dropped from over 30 million in the mid-1940s to about 14 million in 1975.

Herding sheep, Imperial Valley, California, 1961.

Checking quality and length of fleece to determine if the flock is ready for shearing,
New Mexico, 1957.

After the war the poultry business was changing from small family flocks to giant automated "factory" operations. But no matter what size the operation, when a farmer wanted to raise or sell baby chicks for producing meat or eggs it was crucial to know their sex; the males or cockerels, weren't much good for anything. So trained "chick sexors" like this one from Illinois, who traveled a certain territory, would be hired; and in 1947 these people played an important role in those transition times. As poultry raising became bigger and more consolidated, some operations took on permanent employees for this tedious work. Using techniques imported from Japan in the 1930s, the trained chick sexor picks up a day old chick, inspects its sex organs, and tosses it to either the male or female side of the table at the rate of about 1,000 an hour with 95 percent accuracy. The cockerels are later destroyed.

A hired irrigator with his shovel on a large California farm directs water into an asparagus field in 1962. In the U.S. by 1975 there were more than 27 million acres of crops, 8 million acres of hay, and 4.5 million acres of pasture that were irrigated, partly by the pumping from underground acquifers, and also at low subsidized prices through the aqueducts and ditches of publicly constructed water storage and delivery systems.

Bringing sick calf to barn,
Virginia, 1951.

In 1951, Dr. Saylor, a veterinary near Canal Winchester, Ohio was called about a dairy cow with a lodged calf at birth. Saylor tied a rope to the calf's legs, and helped guide the twisted passage while the farmer and a neighbor pulled.

Most farmers with livestock have some knowledge of preventive medicine, but there are situations when a professional is needed. In 1975, at the close of the central period covered by this book's photographs, there were 29,000 veterinarians serving the U.S. farmers. Keeping livestock and poultry healthy was costing about $400 million a year.

An Alaskan farm girl feeding the family chickens near the town of Palmer, 1947.

In 1947, to prepare the land for farming in the Matanuska Valley near Anchorage, Alaska, trees were cut, bulldozed into windrows and burned. Frank Balogh, a homesteader-adventurer-farmer, and his family, transplanted from Ohio, were clearing the last of the snags before burning when this photograph was taken. Next season the land was worked with a plow and disc, and bore its first crop the following season. The Baloghs were building a log cabin at the time and had already had a bear poke its head in the kitchen door. This event caused the immediate posting of a loaded rifle over the mantel.

Improvised bottle-feeding of dairy calves when first weaned from their mother, on the Orval Proud dairy, Miles of Farms, Michigan, 1947.

Along the Mile of Farms in southern Michigan in 1947 a farmer heads for early morning milking in the newly-built barn.

In 1948 this Ohio farmer was one of the first in his area to install an automated barn cleaner. A motor-driven belt with attached paddles pulled manure to the gathering pit, and on to the spreader outside the barn.

The Howard Raffety farm family at supper on their farm near Grinnell, Iowa, 1975.

Mechanization and genetic progress helped ease many unpleasant farm chores, especially at harvest time. For example, machines harvested specially bred, firm-skinned canning and catsup tomatoes that ripened all at once and could withstand rough mechanical handling. But table tomatoes still must be hand harvested, often by Mexican laborers who are not opposed to stooping; Imperial Valley, California, 1960.

Farm field work at dusk, Ohio, 1951.

Technology

*A*dvances in chemicals, communications, computers, and genetics—along with mechanization—were key factors in the changing agriculture. In 1962, U.S. farmers were purchasing about $3 billion worth of farm tractors, power attachments and other machinery each year.

Immediately after World War II, there were still many examples of old-fashioned farming techniques such as this wheat threshing near Mansfield, Ohio in 1948. Such operations were soon replaced by highly organized professional harvesting crews moving with the season from south to north with teams of huge self-propelled combines, through the ripening wheat belt.

At the Bud Antle farms near Salinas, California in 1963, a lettuce field-packing rig enabled twenty-six workers to do the work of forty-five. The lettuce is cut, outer leaves stripped, the heads poly-wrapped and boxed for shipping.

A mechanized potato harvester at Zuckerman Farms near Stockton, California, in 1963, had cut the cost of digging spuds from 17 cents per hundred pounds to 8 cents. The machine and crew of eight replaced thirty human diggers. Lumbering through the hot, dusty fields, the machine's subsurface blade dug up the entire plant, processed it through the traveling shaker to dislodge the potatoes, moved them past human sorters and on to a conveyor belt that dumped into a truck moving alongside. Two workers pick up potatoes the machine has missed.

Another example of an innovative mobile picking-to-packing factory in the field for carrots, designed and operated at the Bud Antle farms in 1963. A plow loosens the carrots, workers snap off the tops and put the carrots on a conveyor belt for cleaning and washing there in the field. Women workers sort them out for polyethylene bagging and boxing. Jumbo carrots are bagged on the side for restaurants and institutions. This unit and its crew of seventy could harvest and pack three railroad carloads in a day. Similar machines, only slightly modified, for harvesting carrots, potatoes, lettuce, canning tomatoes, grapes and orchard crops are still in use today.

Mechanical harvest near Bakersfield, California, 1961.
Each machine can pick as much in an hour as a handpicker can in seventy-two hours.

By the mid-1960s, almost all cotton was picked by
machine but there was still hand-picking in some areas of
the South, such as this Texas farm in 1960.

A more modest, straight line, self-propelled irrigation system leaves fences and fields festooned with icicles after an unexpected freeze, Utah, 1975.

Left and Right: Automated, self-propelled irrigation systems were developed after World War II. Their circular patterns are now common sights from the air, stretching from the Corn Belt throughout the West, as is this one in a potato field near Eureka, Nevada in 1963. From a pump in the center post, the water is pressured into the nozzles of the rig, which circles, unattended, soaking 160 acres every twenty-four hours.

Plant geneticists worked with major crops before and during the post-war years—especially through the laborious and complex breeding process of hybridization—to gain plant vigor, uniformity, and increase yields and resistance to disease and insects. This was particularly effective with corn. Research also produced hybrid seed for sorghum—sometimes called milo—a feed grain grown mostly in dry areas such as western Kansas and Nebraska. In this 1959 photograph a hybrid sorghum seed field is striped with "male" and "female" rows. The female variety has been made genetically male-sterile, so pollen from a selected different variety in the adjacent male plant rows will drift onto the female rows, producing the hybrid cross. Male rows will be harvested as routine feed grain; only the darker-toned female rows will be harvested for seed.

Weighing is a ubiquitous activity on farms and ranches. A piglet is weighed on a small Indiana farm in 1955.

Computers became an important farm tool during the post-war years. In this 1960 photograph on the Codding spread, a large Oklahoma cow/calf breeding ranch, beef cattle weight was recorded directly into a computer.

A wheat sample is weighed for grading at a local farm elevator near Kingston, Ohio in 1949.

*U*nexpected freezing can decimate yields from fruit and nut orchards, so growers developed various types of smudge pots and air fans to raise temperatures a few degrees by keeping warm air circulating with the settling cold air. Environmental concerns such as noise and air quality have rendered almost obsolete the two methods shown here, now largely replaced with electric-powered fans.

A ram-jet rotor stirs a warm breeze in an almond orchard near Ripon, California, 1962. Adapted from helicopter technology, a 24-foot stainless-steel rotor blade, powered by a small ram-jet on each tip, is mounted on a 38-foot shaft, which also serves as a 1,000-gallon fuel tank. The ram-jets heat the air while they power the fanning blades. One machine will service 20 acres at about half the fuel costs for the 400 smudge pots that would be needed for the same job. Even so, the high initial cost, along with the noise factor, limited widespread acceptance of these units.

Lighting old-fashioned kerosene smudge pot, near Modesto, California, 1961.

*D*uring this era there was a great increase in the on-farm use of chemical fertilizers—especially nitrogen—as well as numerous pesticides, fungicides, herbicides and defoliants; from around 14 lbs. per acre in 1950 to 70 lbs. in 1970. Farmers were pleased with greater operational efficiency, along with economic benefits of greater yields. But there were offsets. Insects developed resistance to specific poisons. And there was an awakening public concern leading to government action to control toxic buildups in the water and soil.

Hand spraying insecticide on vegetables on small Wisconsin farm, 1947.

Herbicide spraying of rice fields for weed control, Colusa,
California, 1961.

Mechanized spraying of orchard for insect control near
Porterville, California, 1962.

Artificial insemination can easily introduce a variety of bull sires within a farmer's cow herd, and conception is more controllable. Today's common practice of artificial insemination began in the 1950s. It is widely used in dairy herds where the estrus cycle can be observed on a day-to-day basis. In this 1964 photograph on a large dairy farm near Modesto, California the contract inseminator is feeling for the receptor spot in the cow's uterus before inserting the vial of sperm. Artificial insemination also is now used occasionally in relatively small, high-quality beef herds that are maintained primarily for breeding stock. Natural breeding still is the most practical for large cow/calf beef herds on open ranges in the West where animals may be observed by the rancher only a couple of times a year.

Separating cream in kitchen after morning milking on upper Michigan dairy farm, 1946.

A superior Hereford herd sire, Ohio, 1948. Beef bulls such as this, roaming freely with the cows, are generally calm and gentle. In comparison, confined dairy-breed bulls tend to become irritable and mean-spirited under rigidly controlled breeding conditions. Dairy bulls are often used for nothing more than semen collections for future artificial insemination.

In the three decades after World War II, many complex techniques were developed to control weed seeds, insects, and diseases on land used for high-income crops— such as on this strawberry farm near Salinas, California in 1963. A powerful liquid chemical fumigant was injected by tractor-drawn machines in alternating strips six inches under the soil, then covered with 11-foot-wide polyethylene tarpaulin strips to keep the fumigant from evaporating. After forty-eight hours, the strips were removed and the costly operation repeated on the alternate rows.

*R*esearch into the nutritional value of livestock feed components was a part of the new technology during this era. By using synthetic protein such as urea, it became possible to increase the quantities of cellulose roughages such as ground corncobs, stalks, almond hulls, and others, depending on geographic location, in proportion to such staple nutritional ingredients as corn, barley, and other grains.

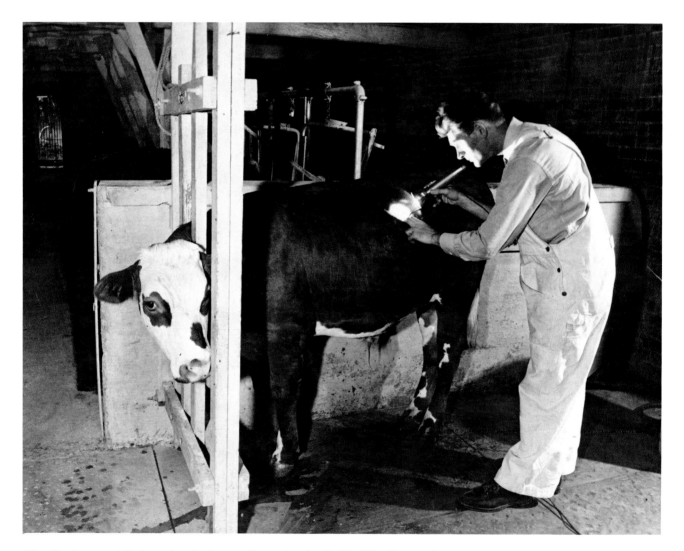

After feeding specially formulated mixtures, livestock scientist Dr. Wise Burroughs was preparing to take a sample for microscopic examination of bacterial action from the first stomach, or rumen, of a steer after removing a large cork from a surgically produced permanent opening in the rumen wall, Ohio, 1948.

Post-war development of huge cattle feedlots for off-the-range yearlings helped flatten the market cycle. Rations for cattle in big feedlots like this one in California in 1965 were already being formulated by computer for proper balance of carbohydrates, fats, proteins, minerals, fibre and vitamins. As the costs of ingredients change, feed rations can be rapidly altered to create least-cost mixtures. Feedlots can hold 25,000 or more head at one time, feeding into troughs from automated, self-unloading wagons. The annual crop may be over 50,000 head, gaining in weight around 300 lbs. per animal in 150 days.

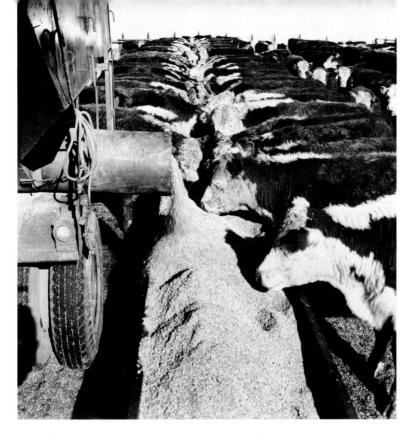

Iowa feedlot, 1952.

In the early 1950s, rising farm labor costs caused fruit and nut orchardists to develop harvesting machines like this prune tree shaker. (To an orchardist, a "prune" is a plum that will be dried.) A long, tractor-powered steel arm with a rubber-cushioned claw grips the tree and shakes the fruit out of it in a matter of seconds. The prunes fall into a self-propelled canvas catch frame and roll into bins; Mills Orchards, Hamilton City, California, 1962.

Other crops—like almonds and walnuts—were also shaken from trees onto leveled ground, then gathered for processing by machines resembling huge, old-fashioned carpet sweepers. In this photograph near Porterville, California, in 1962, walnuts are being emptied into bins. Hanging dead crow was believed to frighten away other crows.

Even with increased mechanization, human decisions remained important. In this western wheatfield in 1954 the farm manager signals harvest combine operator as to which row should be the first or "opening" cut.

Haying on an Ohio farm in 1953.

Land

*O*f the roughly 2.3 billion acres of land in the United States, about one-third is forested. More than half is used for livestock and crops. But each year about a quarter-million acres of cropland are taken away for urban development, right-of-ways, highways and airports.

In addition to sun and soil, plants need water to grow. To provide moisture under arid conditions there were many advances in irrigation equipment during the post-war years. Even so, many western farms were still hand-irrigated by workers with shovels who moved through the fields like this one in California in 1958, channeling the water down the rows from the perimeter ditches.

Burma Shave advertising signs, like this one photographed after an Indiana sleet storm in 1947, are nostalgic symbols of a bygone era on the land. A series of six small signs posted close to rural roadsides tersely recounted whimsical poems. They gradually disappeared in the post-war years as freeways took over and traffic regulations pushed all signs back from the right-of-way.

Foggy morning on Connecticut farm, 1951.

Abandoned barn, Ohio, 1946.

Not all farms shared in agriculture's growth and prosperity at the end of World War II. This recently widowed farmer in central Ohio in 1946 still lived on his dilapidated old farm with his children, but worked in town toward the day he could save enough to move off the farm permanently. Another farm would soon be gone, but the land and future production potential would remain.

Transporting water from northern to southern California, with diversions for irrigation, the California Aqueduct slices through the flat farm land of the state's Central Valley near Bakersfield, 1966.

In the rolling terrain of Knox County in central Ohio in 1950 this farm was plowed on lines contoured to follow the slopes, and keep rain gullies from forming. Not yet using a self-propelled corn harvester, this farmer stacked the crop in drying shocks, awaiting pickup for shelling.

Snow provides a blanket and needed moisture for the coming spring. Sub-zero temperatures kill some overwintering insects. Winter may also be a time for the farmer to sleep a bit later in the morning. Two winter scenes in central New York State, 1948.

In the early 1960s, some public lands were opened to "desert entry homesteading."
This farmer takes a drink from the irrigation ditch on his homestead potato farm near
Eureka, Nevada in 1963.

So-called "dryland" farming in Utah in 1947. These farmers tended field work themselves, but most irrigation was soon to be handled by foreign-born hired irrigators.

Very young tomato plants being started under weather-protective paper covers.
Matanuska Valley, Alaska, June 1947.

End of day's plowing on large Montana ranch, 1960. ▶

A last minute stray has been brought in for the herd milking on a dairy farm near Petaluma, California in 1962.

Mount Vernon, Ohio, county seat for Knox County, in the heart of Ohio's farming region, 1950.

Rural land has many different looks. In south-central Wisconsin in 1947, horse-drawn hay wagons gather the windrowed crop.

Eroded grazing fields, formerly farm woodlots that were logged and converted into pasture. Northern Michigan, 1946.

Bromegrass in pasture, Ohio, 1946.

Sorghum heads, Nebraska, 1959.

Iowa farm soil, 1975.

Corn plant, Iowa, 1950. The tassel at top of plant is the male part. It sheds pollen during a two week period in July. The tiny pollen grains drift down to the waiting silks, the female part, fluffing out the end of the forming ear. One grain of pollen on one strand of silk equals one kernel of corn on the ear at harvest time.

Seed onion field, Utah, 1947.

Wine grape delivery, near Sonoma, California, 1975.

*I*n the early 1960s mechanical grape harvesters were developed, mostly for raisin and juice varieties. Grapes intended for mechanical picking are trained along a trellis, and have long stems for clear access to the cutting blade. By the late 1980s about half of all grapes were being harvested mechanically, while the other half—those grown on hillside areas—were being harvested by hand.

Grapes being harvested by hand in California's Napa Valley in 1961.

Vineyard workers in spring, pruning and tying vines, Napa Valley, California, 1957.

Farmer James Quisenbury and purebred Angus bull, Kentucky, 1948.

Livestock

*H*usbandry techniques have varied from farm to farm and region to region. During the three post-war decades livestock tended to shift to larger animal numbers—on fewer farms. Alert farmers were better able to anticipate price trends in cyclical changes in animal numbers.

Farmer moves the piglets from the sow to the feeding pens.

Very young piglets may eat ten or twelve meals a day.

Holstein dairy calves shortly after weaning, California, 1960.

When World War II ended there were a number of popular dairy cow breeds: Holstein, Guernsey, Brown Swiss, Milking Shorthorn, Jersey and others, each with certain perceived advantages. The large, rangy, black and white Holsteins, giving the most milk with the lowest fat content, became by far the dominant breed. In 1975 the average dairy herd across the U.S. still numbered only about twenty-five head. But there had been a big move toward larger herds, particularly in California where 500 to 1,000 head on a single highly mechanized dairy was not uncommon.

Guernsey cow, Illinois, 1946.

Dairy cow slips on wet floor of barn. Mile of Farms, Michigan, 1947.

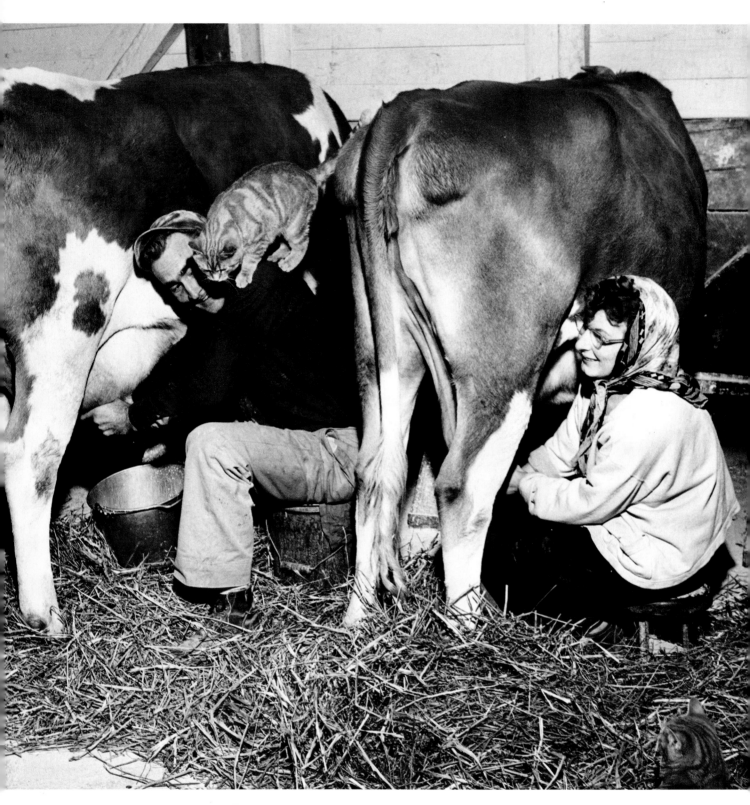

Dairying is one of the most demanding types of farming. Cows must be milked twice a day, 365 days a year; they can't be shut off for a vacation. Buildings and equipment must meet rigorous health standards. Labor to help with dairying was hard to get in post-war years, so dairying was largely a family operation. A few dairy farmers, such as this couple near Wooster, Ohio, in 1952, still felt that hand milking, with its more personal contact with the cows, was better than machines.

Farm cats getting milk bucket drippings.

By the end of World War II the work-horses that had powered American agriculture—the ponderous Belgians, Percherons, and Clydesdales—had long since given way to truck and tractor. But some farms like the Bonnie Brae Farms, near Wellington, Ohio, in 1954, that had expertise in raising quality workhorses from the past, had switched to raising horses for riding and racing. The animals in these photographs from Bonnie Brae are a breed known as Standardbred. Used with a two-wheel sulky, the best of these trotters and pacers traverse a mile in very close to two minutes. They are descended since 1851 from the great Standardbred stud, Hambletonian, who begat 1,331 foals in his 24 years of service.

Part of the Bonnie Brae herd.

Standardbred mare and foal at Bonnie Brae farms.

A shipload of sheep from Australia unloads at the port of San Francisco in 1960.
These were quality-bred animals ordered by a U.S. sheep raiser for improving his
existing flock. Today, shipment of cattle, hogs, and other livestock is usually in test
tubes in the form of frozen semen or embryos, and usually by chartered jet aircraft.

A New Mexico shepherd in 1965 moves part of his flock from one pasture to another, happening to cross the plaza in front of the St. Francis of Assisi Church near Rancho de Taos. Shepherding on the range is a lonely life, and wandering dogs and wild predators can play havoc with flocks, especially at lambing time. Most hired shepherds are Basque or Peruvian. They may spend six months virtually alone with their flock.

With a plentiful supply of loose feed and open shelter, rats always have been terrible pests around the farm. As long ago as 1948 when this photograph was taken, the nationwide farm rat bill was an estimated $300 million in lost or contaminated grain and other produce. The poultry house, like this high-density shed with open grain feeders in Michigan in 1946, is a special target for rats. Three rats will eat as much as two adult hens, and if the feed runs out they may run amuck among the baby chicks.

Above: laying hens raised in cages, Michigan, 1946. Right: young chicks, Missouri, 1960.

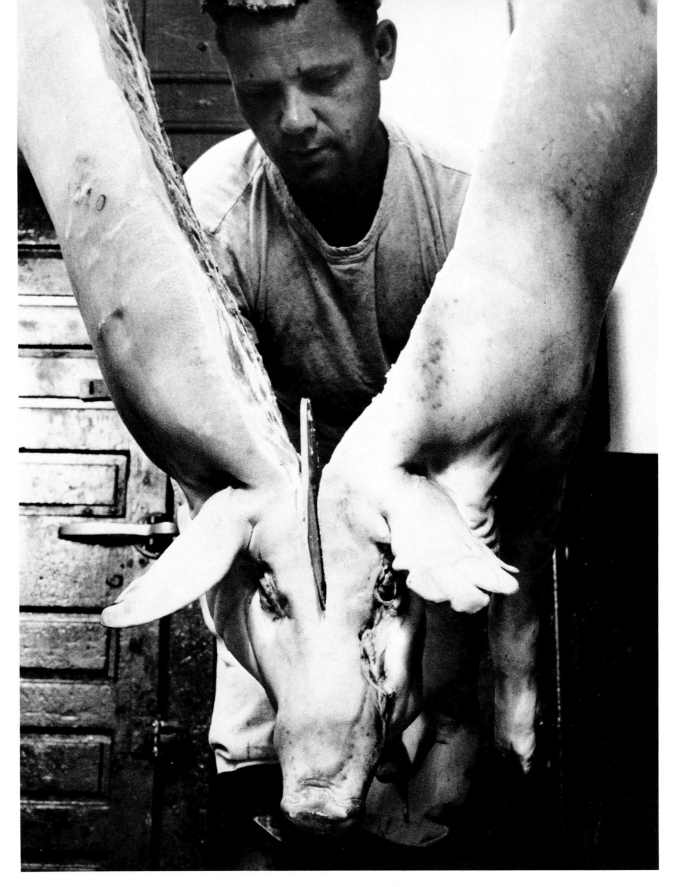

This photograph was made at a rural custom slaughterhouse in Indiana in 1955. The cycles of living, and dying with crops and livestock are keenly seen and felt as an everyday occurrence on the farm. Yet most urban consumers probably are more aware of food prices and nutritional values. Thirty-five percent of the calories consumed in the U.S. in 1975 came from animal fats, compared to 13 percent in Brazil and 6 percent in India. Public preferences had been changing. During the three decades after World War II, lean-cut pork chops, hams and loins increased from 37 percent to 55 percent of hog carcass by 1975.

The Border Collie is one of the best breeds of dogs for helping the farmer or rancher handle livestock. These dogs—medium size, wiry and usually with black and white coat—are bred with a natural inclination to herd sheep or cows. Some farmers swear that these Collie dogs, after some weeks of training, can do everything about herding livestock during the day and bringing them in at night—everything except lock the barn door. Near Gambier, Ohio in 1946, a sheep farmer and his Border Collie pause at dawn at the field gate to savor the sounds and smell before starting the day's chores.

Cattle roundup, cowboy heading off strays, 1957.

Continuity Amidst Change

*O*n the ranches in the western United States, from Texas to Montana and west to the Pacific, the raising of range cattle in the great cow-calf herds is a classic example of continuity amidst change. Depicted in countless movies, songs and writing, the tradition of western cattle raising began generally in the mid-1800s. The Indians were subdued. The buffalo was hunted to extinction, leaving land and grass for the fledgling cattle industry. First came the tough Texas Longhorn cattle that could subsist with little water and on sparse range during the long drives to the few scattered railheads for shipment east. The Longhorns were gradually replaced, and the gentler, meatier, more palatable breeds like the Herefords from England and the eastern United States began to dominate the range, because they were able to more efficiently use the good grass that grew over the vast Plains areas.

The genetically more advanced breeds needed water supplies more closely spaced than did the buffalo and hardy Longhorns. The lowly windmill, powering a small waterwell pump with its adjacent drinking tank, became an important fixture for the cattle industry on the long reaches of the unelectrified, semi-arid, western rangeland.

A region-wide system of branding—burning registered identification insignia into the hides of very young cattle—was almost universally used beginning about the mid-1800s. This survives and remains an important tool in managing the wandering, sometimes intermingling, herds over far-flung ranges.

Many ranches were, and still are, huge. But by the late 1800s the homesteading laws, spreading railroad lines and a mundane but crucial item, the newly invented barbed wire fencing, gradually brought about ranches of more moderate size, and the end of the "open" range. The open range was partially replaced by public lands where large herds of range cattle could graze under a fee-permit system.

Most of the photographs in the following set were taken in the mid-1960s at the 900,000 acre 3V ranch near Seligman, Arizona, 200 miles northwest of Phoenix. At elevations of 4,000 to 6,000 feet, the ranch stretched from Old Route 66 to the southern rim of the southern Grand Canyon. The sparse range provided grazing for an average of only one cow per 80 acres.

Every autumn the cowboys on horseback, along with a pickup truck serving as a "chuckwagon" carrying spare saddles, bedrolls, camp gear and food, headed for shacks and corrals of the "line camp" nearest to wherever the herd had wandered. The roundup could last for a couple of weeks or more and was known as living "on the wagon," a colloquialism derived from the unwritten law that there were no hard-alcohol beverages allowed on the cattle roundups.

In 1965 there were about 3,000 animals rounded up and brought to the railroad loading ramp at one of the half-dozen corral units spread over the ranch. During the operation the cattle were counted, new calves branded, male calves castrated to become steers, and the herd vaccinated and dipped for insects. About 500 heifers and 300 very young calves and a few selected bulls were turned back to the open range with the foundation herd of about 1,500 brood cows to produce next year's calf crop. Roughly 700 steer calves were sold and shipped for more rapid fattening to large feedlots, or to smaller farms or ranches, usually in the Corn Belt, for further pasturing and fattening. Some older animals were culled for sale to meatpackers.

In the minds of many urban dwellers, the cowboys doing this work remain a romanticized "special breed." They are a proud lot; it is a tough demanding life. Cowboys know how to let off steam in their off-work time. The many rodeos around the nation are showcases for their daily work. When working the cattle they may use pickup trucks, modern vaccination serums, burn butane in the chuckwagon stove, and heat the branding iron with a kerosene torch; but cowboys, even today, live and work pretty much according to the code and methods established when men first came to the open range. ■

Cowboy ropes a horse from the remuda for the day's roundup work, Seligman, Arizona, 1965.

Beef cattle being put through insect
and disease "dip" trough near
Seligman, Arizona, 1963.

Ranch hand carrying a sick calf,
Seligman, Arizona, 1963.

Cattle herd crossing American River, California, 1957.

Ranch hands rolling out in the morning, Arizona, 1963.

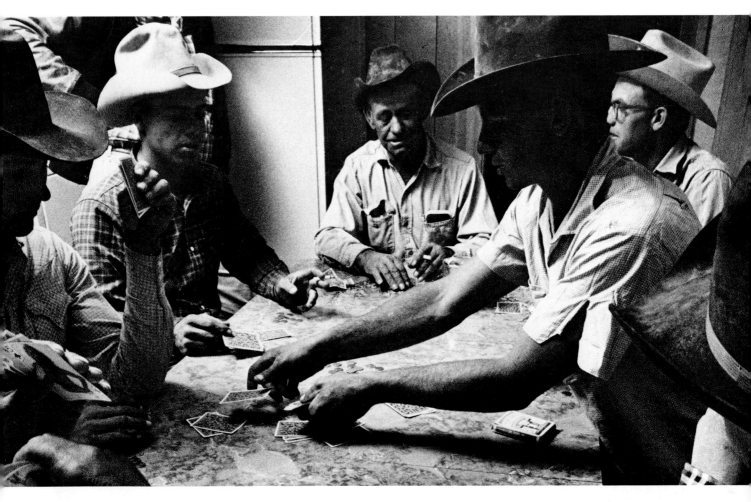

Evening poker game in the line shack during cattle roundup, Arizona, 1963.

Dawn breakfast, cattle roundup, Arizona, 1963.

Working cowboys on roundup near Seligman, Arizona, 1964.

Cattle roundup in the high country during winter near Seligman, Arizona, 1963.

Barbed wire, windmills and cowboys; three major
elements in changing the face of the open range.

Mr. and Mrs. Len Cole owned 90 acres on the Mile of Farms in Michigan, 1947.

People

Despite improvements in technology the great strength of America's agriculture is still her people.

In the years after World War II farm families lived, worked and played together. But living patterns were being changed by advancing technology. This was particularly true among young people. During this period radios and air conditioning appeared in tractor cabs, and computers began keeping records. But fewer young people were taking up farming.

Sheep rancher, Judd McKnight, center foreground, with his family near Roswell, New Mexico, 1957.

Farm family, central Illinois, 1948.

Laverne Raffety, seventy-five years old, tried to convey some of the wisdom of his experience and enthusiasm for the fine points of farming to Michael, then twelve years old.

The next day Michael waits for the school bus. Michael helped around the farm and liked to run the farm machinery, but as time passed he became more interested in computers than in farming. After college he went to work for a currency brokerage in Chicago. His father, Howard (Laverne's youngest son), wonders whether the farm can continue to be family operated after he and his wife, Suzy, have retired.

Even with advancing technological conveniences and a narrowing of differences between urban and rural lifestyles, keeping young people on the farm was a growing challenge. On their 300-hog farm near Grinnell, Iowa, in 1975, Laverne Raffety, back to camera, and his grandson, Michael, were counting litters of young pigs.

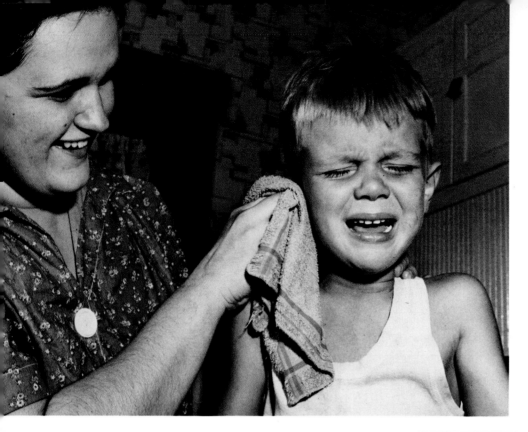

On a Saturday morning in 1946 on a farm near Laurel, Indiana, a young lad protests his mother's efforts to get him ready for a big event—the weekly shopping and movie trip to town.

Not all farm living involved sweat and toil. In southern Ohio in 1947, father and son head for the farm woodlot to do a bit of crow hunting.

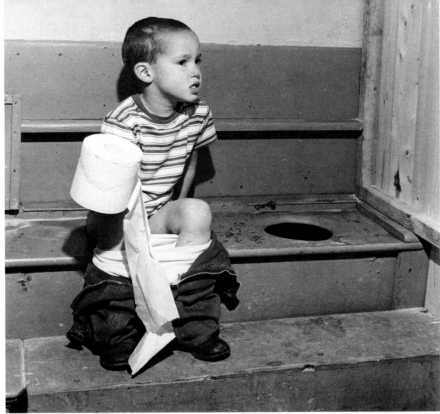

Connecticut farm, 1951.

Mrs. Wagner's Rolling Store made regular stops in 1953 on its route through a part of rural Kentucky in which some farm families either did not own autos, or did not trust them to get to town and back for shopping. In this photograph a customer approaches the converted school bus which carriers notions, simple medicines, clothing and household items, and Mrs. Wagner would take orders for next week's delivery.

Rural electrification and propane gas services were almost universal by the end of World War II, but many farms like this one in Indiana in 1947 still did not have a wide array of such appliances as a clothes dryer. Besides, even though long-johns may stiffen on a frosty day, many farm wives simply like the freshness of air-dried laundry.

▲
The Wimodausians, a farm women's literary society, held their monthly meetings in an Ohio farmhouse in 1950. Women adopted cultural activities to rural life by forming study clubs, gardening clubs, and bridge clubs, as well as organizing musical and dramatic entertainment. Their desire for self and family improvement, and sociability remain at the heart of small town life even today.

◄ Summer farm work was the reason for the tradition of winter school and summer "vacations," beginning in the days when about half the nation's population was engaged in farming. These youngsters are returning from school in the western farming community of Hurricane, Utah in 1947. Streetside gutters (lower right) in this semiarid region sluiced precious irrigation water to fields across town.

Rugs rolled up for Halloween square dancing on Connecticut farm, 1951.

Farm women shucking peas for supper, southern Virginia, 1951.

Grandmother and granddaughters in farmhouse living room, New York State, 1950.

Stanley Smith married the granddaughter of Eliphalet Williams, to whom Isaac Fulton had long ago sold the first piece of settled land on The Mile of Farms (see text) in southern Michigan. Smith eventually sold that farm to bustling, high-tech Omega Farms, today's largest landowner on The Mile. But when this picture was taken in 1947 Stanley Smith, 49 years old, was a friendly philosophical farmer who enjoyed the slower rhythms of pastoral life. He grossed about half the earning potential of his land, and went hunting with his dogs when the spirit moved him. For Smith, the raising of sheep and a few dairy cows, along with farming some field crops, was a way to pursue the good life. Yet, mindful of the bank failures of 1933, Smith put aside a bin of wheat each year as a "bank" for paying his taxes.

In the next thirty years, many became more serious about their farming, such as Harry Jacobsen and his son, planting corn in Iowa in 1975.

(See on following page) During the thirty-year period after World War II, wages and working conditions improved for farm labor that was mostly migratory, working in such crops as fruits, nuts and vegetables. But farm labor/management improvements came gradually, sometimes as a result of an enlightening political atmosphere, but often as a result of disputes that led to the rise of farm labor unions with leaders like Cesar Chavez, who called for consumer boycotts and strikes.

Strike in progress at a tomato field in the central valley of California in 1960.

Farm life was, and is, demanding for women. When not doing housework or caring for children, they are often outdoors helping their husbands. And like their husbands, they worry about the future, Champaign County, Illinois, 1948.

In the Corn Belt and other grain growing areas some of that worry was, and is, linked to what's happening to prices in the export market. In 1959, in an attempt at better international understanding, an innovative farmer and hybrid seed corn marketer, Roswell Garst (center) with guest Adlai Stevenson (left) hosted Soviet Premier Nikita Khruschev at the Garst farm and seed production plant at Coon Rapids, Iowa. Soviet agriculturists made exploratory trips to U.S. farms, and Garst led reciprocating visits to Russia, in both an advisory capacity, and to sell seed corn and machinery to the Soviets. Garst, who died in 1977, felt strongly that food and hunger knew no boundary lines. He liked to point out that the southern edge of the Russian grain belt is at the same latitude as Minneapolis, and he wondered what the status of American agriculture would be if we grew all of our grain north of that point.

Despite newspapers, magazines, radio, television and movies there still was a gap in communication and understanding between the farmer, city, and suburban dwellers. Some say it resulted from differences in their daily life, although in truth they were living more alike than ever before. Part of the early social structure of The Mile of Farms in southern Michigan in 1947, was the one room schoolhouse. It closed in 1957, being consolidated with a nearby town school at Webberville. The old building was used as a basketball court for the neighborhood until it was torn down in 1985.

In 1947, on The Mile of Farms, Margaret Suttell was the teacher at the one room school until it was closed. She mothered, scolded and instructed twenty-three pupils in eight grades each year. Miss Suttell once wanted to be a barnstorming aviator. She died in 1985, the year the school building was torn down. "Country schools turned out better pupils than city schools," she claimed, "because the youngsters received more personal attention."

The "top hand" cowboy manager on the 3V Ranch near Seligman, Arizona in 1965. His wife served as cook on roundups.

Cajun farmer, Joe Stelly, age 42, and his wife posed on their farmhouse step near Opelousas, Louisiana in 1949. (Cajuns are descendants of Acadians who were driven out of Canada by the British and found refuge in Louisiana in 1765.) The Stelly's owned seventy-five acres in the swampy Bayou Teche country. The main crop on their neat farm was sweet potatoes with a bit of cotton, corn, some hogs, chickens, and a few scrub cattle. Mrs. Stelly was noted in the neighborhood for the intense quality of the garlic she grew in her garden. Joe had no tractor; but in that area they were coming fast. He still did all the field work with two mules and a horse. A sharecropper family lived on Stelly's farm in a simple dwelling. He used Joe's tools and mules and got one-third of the cotton and potato crop on about thirty acres; if lucky, he grossed $1,200 a year. It was said that the Cajun husband was encouraged by his wife to believe that he was head of the household. She stands behind him, but has both hands against his back, pushing.

Among the early groups of Mexican laborers beginning to flood into the southern
states, this worker was cross-breeding flower varieties at the Burpee Seed Company
fields near Lompoc, California in 1956.

This young Tennessee farm couple in 1952 picked cotton by hand, ▶
and had just emptied their shoulder-drag field bags in the storeroom.

Many aspects of farm life were changing, but most farm families still cherished their
land and made it fruitful by working hard, while enjoying themselves at the same time.

Acknowledgements

By Joe Munroe

My first farm photography assignment was for *The Farm Quarterly* on the J.F. Walker sheep farm near Gambier, Ohio, in 1946. To a "city boy" from Detroit it was a real learning experience. That visit resulted in a lasting friendship between my family and the Walkers, as has happened with many of the several hundred other farms, ranches, and rural families I've visited over the years who have been so helpful.

For my experiences in agricultural photography during the three decades after World War II, I am indebted to the staffs of what were then the three major national farm magazines: *Successful Farming*, *Farm Journal*, and *The Farm Quarterly*. In the case of the first two I note with thanks the help of Jim Roe, Austin Russell, James Borcherding, Carroll Streeter, Max Kille, and Bob Fowler. It was *The Farm Quarterly* that gave me my start in photojournalism during that period. Until that publication's demise in 1972, I was especially affected and enlightened by Ralph McGinnis, Fred Knoop, Aron Mathieu, George Laycock, Charles Koch, Tom Huheey, and most particularly by Grant Cannon, with whom I had a friendly and close working relationship for many years.

In non-farm magazines, Seville Osborn-Johnson, Picture Editor at *Fortune* in the '50s and '60s aided me in what became a major *Fortune* photographic essay on farming innovations.

In the early 1950s I spent some time with the United States Information Agency. I owe much to the encouragement of William F. Bennett and Richard Pollard, then the directors of our staff of field writer-photographers. I later worked with Pollard on *Life* magazine and book assignments that included farm subjects.

There were also agriculture-related corporate assignments. I appreciate very much the encouragement and enthusiasm of Ralph Lyman of Chevron Oil Company, Angus McDougall of International Harvester, Don Fabun of Kaiser Aluminum, Gordon Sears of Kern County Land Company, Roy Dobie of Northrup King Company, Gordon McCleary, Jock Wallace and Steve Moon of Pioneer Hi-Bred International, Inc., and Ray Dankenbring and Robert Kurt of Ralston Purina Company, among others.

In 1947, I became friends with Roswell Garst (1898-1977) and his family at Coon Rapids, Iowa; and their operations were the subject of several of my farm picture assignments. Roswell's son-in-law, Harold Lee, a fine writer, was helpful to me in some of the early planning of this book. Roswell had great energy and knowledge. His love of arguments and his boundless enthusiasm for new thinking—not only in hybrid seed corn, but all of agriculture— were an ongoing inspiration to me. Our acquaintance led to an unforgettable picture assignment for me on the Garst farm in 1959. I was the Garst "family photographer" on that madhouse day that USSR Premier Nikita Khruschev came to visit— along with many dignitaries, friends, and a great crush of several hundred writers, television interviewers and photographers of the world press. (See page 175 for more.)

I also became a good friend of and "official photographer" for Louis Bromfield, the novelist and experimental farmer who died in the late 1950s. His Malabar Farm was not far from our home in Mount Vernon, Ohio, in the late 1940s. Like Garst, but in different directions, Bromfield was a provocative thinker, and was helpful to me in those early days. Several pictures from the Garst and Bromfield farms are in the book. See pages 9, 63, 86, 107 and 175.

In writing the captions, I relied heavily on my own notes and memories, as well as information gathered by writers and editors in the accompanying articles. Some information also was gleaned from various editions of the U. S. Department of Agriculture's Yearbook of Agriculture.

Design and layout is crucial in a picture book, and I wish to very much thank Larry Weber for his help to me in the early planning stages of this book, and to George Myhervold for carrying the original concepts through to completion.

During those early years I did much traveling, kept weird hours, and often came home reeking of barnyard smells. My wife, while becoming an artist and printmaker in her own right, has raised our four children and fussed over nine grandchildren. For her patience and encouragement during those early years of my free-lancing, the biggest thank you of all goes to Virginia. ■

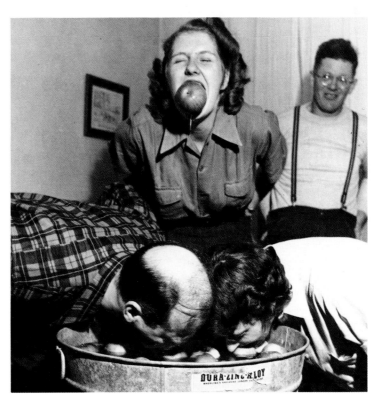

Photographs can picture fun on the farm as well as work. Halloween on a Connecticut farm, 1951.

Acknowledgements and Suggested Further Reading

By Kirby Moulton

Numerous people have been very helpful in the preparation of this book. Several should be mentioned from more recent times.

Foremost is Professor Tom Schaub from the University of Wisconsin, who provided wise assistance in bringing our text and captions into a more understandable form. Tim Wallace, an economist and colleague at the University of California, was instrumental in helping hone some of the concepts and policy issues during several stimulating discussions. Other ideas were drawn from discussions with Jerry Siebert, John Mamer, George Goldman, Howard Rosenberg, Bees Butler, Robert Curley and Calvin Qualset, all of the University of California staff.

Myiesha Bradford was of great help in preparing the manuscript and helping illuminate the mysteries of word processing.

Printed sources that were most helpful to me, and would be good reading for those wanting to explore the development of American agriculture and the issues of the post-war period, include the following. Willard Cochrane's *The Development of American Agriculture, an Historical Analysis* (University of Minnesota Press, 1979) puts farming in a long-term perspective and is particularly fascinating in the discussion of technological change, supply and demand, and the "treadmill."

Roosevelt's Farmer (Columbia University Press, 1961) contributes a highly personal account of the conditions and politics of American agriculture before the Second World War. This book, by Dean Albertson, describes the base that nurtured the subsequent revolutions in agriculture. Milton C. Hallberg presents an objective and well-documented analysis of changes in U.S. agriculture between 1950 and 1987 in his paper *The U.S. Agricultural and Food System: A Postwar Historical Perspective* (Regional Center for Rural Development, Pennsylvania State University, 1988). Gordon Fite examines changes in agriculture, their impact on farms and farmers and the issues arising from them in a very engaging book, *American Farmer's: the New Minority* (University of Indiana Press, 1984).

Two books that focused mostly on California agriculture gave me some insights that were useful in understanding national developments. One book was *A Guidebook to California Agriculture* (University of California Press, 1983), edited by Ann Scheuring. It presents a series of articles on the forces shaping California and U.S. agriculture and is particularly interesting in the sections on marketing and on the consumer. A less widely distributed book is Yvonne Jacobson's *Passing Farms, Enduring Values* (William Kauffman, Inc., Los Altos, Calif. 1984) which, as an individual reflection of growing up on a farm in California, sketches the changes going on in agriculture and, particularly, in labor conditions on farms.

Juan Gonzales, Jr.'s *Mexican and Mexican American Farm Workers* (Praeger Publishers, 1985) is an objective look at the migrant labor force that supports much of U.S. agriculture. His chapter on worker typology helps us understand how highly differentiated the labor force is and therefore why labor issues are so complex. Perhaps of broader interest is the U.S. Department of Agriculture publication, *1990 Yearbook of Agriculture: People Working in American Agriculture*. It provides a portrait of the people in U.S. agriculture, what their conditions are and what issues concern them and others.

Books of a more academic vein that provide background for my writing and which can be useful for the motivated reader in understanding policy issues of the post-war period include *Size, Structure and Future of Farms* (Iowa State University Press, 1972), edited by A. Gordon Ball and Earl O. Heady; and *Farm and Food Policies and Their Consequences* (Prentice Hall, 1989) by Kenneth L. Robinson. In the former book, the article by Phillip Raup, "Societal Goals in Farm Size" is particularly good in identifying the ambiguities in our thoughts and policies affecting farm size during the period of concern in my essay. Robinson's book, although written well after the period of primary concern to me, provides a very good description of the evolution of policy, what the issues are, and what alternatives there are.

Luther Tweeten of Ohio State University shared his ideas concerning the impacts of technological change by allowing me to review his 1988 manuscript, *Long Term Viability of U.S. Agriculture*, prepared for the Council for Agricultural Science and Technology. A number of other books have helped develop my perspectives about agriculture, but I think that their influence has been less significant in the writing of this essay than the discussions I've had with friends in and out of agriculture over the past two decades.

The U.S. Department of Agriculture publishes an impressive array of material that describes, classifies and evaluates the complex network that comprises U.S. agriculture. Many of the publications are annual and contain statistical compilations reporting results for several years. There are annual reports on production and prices, by producing regions, for a large number of agricultural commodities. These have allowed us to trace changes in productivity over the post-war years.

Some of the more useful titles include the following: *Chronological Landmarks in American Agriculture; Farm Income Data, an Historical Perspective; Farm Operating and Financial Characteristics; Structure Issues in American Agriculture; Agricultural-Food Policy Review; Annual Report on the Status of Family Farms* (a report to Congress required by law); *Economic Indicators of the Farm Sector* (a series of publications describing economic results); *Food Cost Review, 1989; Farm Production Expenditures;* and *Agricultural Prices*. Additionally, data from the *Agricultural Census*, completed every 5 years, has been useful in tracing important trends in the farm sector.

I hope that truly interested readers will seek Cooperative Extension specialists and will search through farm publications in an effort to better understand the enormous changes now occurring in American agriculture. ■

Notes on Photography

By Joe Munroe

*B*eing interviewed by *The Farm Quarterly* for my first job as a photojournalist, I was asked if I knew anything about farming. With a sinking heart I was getting up to leave while saying, "No." The editor exclaimed, "Good!" In the next twenty-five years I learned about farming, but there was always a sense of wonder and excitement at each new assignment. And with many subjects other than farming, that approach stayed with me.

My early photographic education began with a short course by Ansel Adams in the late 1930s, and thereafter was much shaped by the teachings of Arthur Siegal. I greatly admired the styles of Cartier-Bresson, W. Eugene Smith, Edward Weston, Dorothea Lange and a variety of other photographers, and came to be personally acquainted with some of them.

With my own work I generally took a straight approach, shooting it the way I saw it—as simply as possible.

Photographs in this book were taken with various cameras: Speed Graphic, Rolleiflex, Hasselblad, Nikon and Leica. I prefer hand-holding and natural light whenever possible, but tripod and supplemental flash were used when needed, with occasional special filters and film, such as infra-red emulsions on pages 110 and 111. Films were mostly standard Plux-X, Tri-X and Verichrome Pan; developed in D-76. Time permitting, I did my own processing, and I made all the prints for the engravings in this book. I occasionally used aircraft as shooting platforms, shooting from a helicopter in the air-to-air shot on page 28.

This book of all black-and-white photos reflects the fact that during those three decades after World War II, most farm magazines ran most of their photography in black-and-white. Even so, during that time I did considerable color shooting. A few photographs in this book, because of various unique qualities or subject matter, are black-and-white conversions from color transparencies. They appear on pages 20, 21, 87, 88, 89, 92, 98, 104, and 139.

Some explanation might be of interest about a couple of technical challenges. The shot of Dr. Wise Burroughs peering by fistula into the rumen of a steer (page 106) was made with the camera flash and a second flash synchronized to fire inside the steer's rumen to simulate the light from Burrough's small hand flashlight to shield the animal's tissue from the heat of the old-fashioned flashbulbs of 1948.

For the aerial dusting of the potato field, pages 20 and 21, I was fitted into a plastic clothes bag and strapped, rear-facing, into the converted cockpit dust hopper in front of the pilot of the old Stearman biplane, a military trainer adapted for chemical applications. I was standing atop the half-load of insecticide dust on a makeshift plywood subfloor, so my feet would not get entangled in the whirling augur in the hopper bottom, as the chemical was carried out to the spreading vent under the fuselage.

The hopper normally would have been sealed to keep the dust from being sucked into the slipstream; but the plastic bag with me in it served as the sealing cover, with edges overlapped out and clamped down around the hopper opening. Being backwards during the actual ride was a unique sensation; and it enabled me to shoot back over the pilot to catch the pattern of spewing dust, and the lucky moment of the plane's own shadow.

My objective in photography is for the viewers to feel they are a part of the scene, without sensing the interposition of the camera. I try to combine editorial function with visual excitement—to seize a significant or uncommon moment or design that will intensify the viewers' experience. ∎

Seed corn harvested on the ear being dumped into processing plant hopper, Iowa, 1959.